Modern Methods of Plant Analysis

Editors
H.F. Linskens, Nijmegen/Siena/Amherst
J.F. Jackson, Adelaide

Volume 17

Springer
Berlin
Heidelberg
New York
Barcelona
Budapest
Hong Kong
London
Milan
Paris
Santa Clara
Singapore
Tokyo

Volumes Already Published in this Series:

Volume 1: Cell Components
1985, ISBN 3-540-15822-7

Volume 2: Nuclear Magnetic Resonance
1986, ISBN-3-540-15910-X

Volume 3: Gas Chromatography/ Mass Spectrometry
1986, ISBN 3-540-15911-8

Volume 4: Immunology in Plant Sciences
1986, ISBN 3-540-16842-7

Volume 5: High Performance Liquid Chromatography in Plant Sciences
1987, ISBN 3-540-17243-2

Volume 6: Wine Analysis
1988, ISBN 3-540-18819-3

Volume 7: Beer Analysis
1988, ISBN 3-540-18308-6

Volume 8: Analysis of Nonalcoholic Beverages
1988, ISBN 3-540-18820-7

Volume 9: Gases in Plant and Microbial Cells
1989, ISBN 3-540-18821-5

Volume 10: Plant Fibers
1989, ISBN 3-540-18822-3

Volume 11: Physical Methods in Plant Sciences
1990, ISBN 3-540-50332-3

Volume 12: Essential Oils and Waxes
1991, ISBN 3-540-51915-7

Volume 13: Plant Toxin Analysis
1992, ISBN 3-540-52328-6

Volume 14: Seed Analysis
1992, ISBN 3-540-52737-0

Volume 15: Alkaloids
1994, ISBN 3-540-52738-9

Volume 16: Vegetables and Vegetable Products
1994, ISBN 3-540-55843-8

Volume 17: Plant Cell Wall Analysis
1996, ISBN 3-5430-59406-X

Volume 18: Fruit Analysis
1995, ISBN 3-540-59118-4

Plant Cell Wall Analysis

Edited by
H.F. Linskens and J.F. Jackson

Contributors

A. Bacic L.M. Barone M. Bordenave H. Chanzy
M.A. Coimbra I. Delgadillo K. El-Bahtimy L. Epstein
M. Fèvre M.L. Fishman P.P. Gallego L. Gay E.E. Hood
A.T. Hotchkiss I.M. Morrison W.M.A. Niessen X. Qi
S.M. Read H.A. Schols R.R. Selvendran J. van der Greef
R.A.M. van der Hoeven A.G.J. Voragen V. Vreeland
K.W. Waldron B.P. Wasserman A. Wu

With 36 Figures

Springer

Professor Dr. HANS FERDINAND LINSKENS
Goldberglein 7
D-91056 Erlangen, Germany

Professor Dr. JOHN F. JACKSON
Department of Viticulture, Encology and Horticulture
Waite Agricultural Research Institute
University of Adelaide
Glen Osmond, S.A. 5064
Australia

ISBN-13:978-3-642-64644-7 e-ISBN-13:978-3-642-60989-3
DOI: 10.1007/978-3-642-60989-3

The Library of Congress Card Number 87-659239 (ISSN 0937-8340)

Softcover reprint of the hardcover 1st edition 1996

Typesetting: Best-set Typesetter Ltd., Hong Kong
SPIN 10128397 31/3137-5 4 3 2 1 0 – Printed on acid-free paper

Introduction

Modern Methods of Plant Analysis

When the handbook *Modern Methods of Plant Analysis*, was first introduced in 1954, the considerations were:

1. the dependence of scientific progress in biology on the improvement of existing and the introduction of new methods;
2. the difficulty in finding many new analytical methods in specialized journals which are normally not accessible to experimental plant biologists;
3. the fact that in the methods sections of papers the description of methods is frequently so compact, or even sometimes to incomplete, that it is difficult to reproduce experiments.

These considerations still stand today.

The series was highly successful, seven volumes appearing between 1956 and 1964. Since there is still today a demand for the old series, the publisher has decided to resume publication of *Modern Methods of Plant Analysis*. It is hoped that the New Series will be just as acceptable to those working in plant sciences and related fields as the early volumes undoubtedly were. It is difficult to single out the major reasons for the success of any publication, but we believe that the methods published in the first series were up-to-date at the time and presented in a way that made description, as applied to plant material, complete in itself with little need to consult other publications.

Contribution authors have attempted to follow these guidelines in this New Series of volumes.

Editorial

The earlier series of *Modern Methods of Plant Analysis* was initiated by Michel V. Tracey, at that time in Rothamsted, later in Sydney, and by the late Karl Paech (1910–1955), at that time at Tübingen. The New Series will be edited by Paech's successor H.F. Linskens (Nijmegen, The Netherlands) and John F. Jackson (Adelaide, South Australia). As were the earlier editors, we are convinced "that there is a real need for a collection of reliable up-to-date methods for plant analysis in large areas of applied biology ranging from agriculture and horticultural experiment stations to pharmaceutical and technical institutes concerned with raw material of plant origin".

The recent developments in the fields of plant biotechnology and genetic engineering make it even more important for workers in the plant sciences to become acquainted with the more sophisticated methods, which sometimes come from biochemistry and biophysics, but which also have been developed in commercial firms, pharmaceutical laboratories, non-university research institutes, and medical establishments.

Concept of the New Series

Many methods described in the biochemical, biophysical, and medical literature cannot be applied directly to plant material because of the special cell structure, surrounded by a tough cell wall, and the general lack of knowledge of the specific behaviour of plant raw material during extraction procedures. Therefore all authors of this New Series have been chosen because of their special experience with handling plant material, resulting in the adaptation of methods to problems of plant metabolism.

Nevertheless, each particular material from a plant species may require some modification of the described methods and usual techniques. The methods are described critically, with hints as to their limitations. In general it will be possible to adapt the methods described to the specific needs of the users of this series, however references have been made to the original papers and authors. During the planning of this New Series the editors tried to ensure that the aims and general layout of the contributions are within the general guidelines indicated above, but in addition they tried not to interfere too much with the personal style of each author.

There are several ways of classifying the methods used in modern plant analysis. The first is according to the technological and instrumental progress made over recent years. These aspects were taken into consideration in Volumes 1 to 5 and 11 of this series describing methods in a systematic way according to the basic principles of the methods.

A second classification is according to the plant material that has to undergo analysis. The specific application of the analytical method is determined by the special anatomical, physiological, and biochemical properties of the raw material and the technology used in processing. This classification was used in Volumes 6 to 8, 10, 14, 16 and 18.

A third category is according to the classes of substances present in the plant material and the subject of analytical methods. The latter was used for Volumes 9, 12, 13 and 15 of the series.

Naturally, these three approaches to developments in analytical techniques for plant materials cannot exclude some small overlap and repetition; but careful selection of the authors of individual chapters, according to their expertise and experience with the specific methodological technique, the group of substances to be analyzed, or the plant material which is the subject of chemical and physical analysis, guarantees that recent developments in analytical methodology are described in an optimal way.

Volume Seventeen – Plant Cell Wall Analysis

The first chapter in this volume deals with fractionation of cell wall components. Plant cell walls can be considered to be essentially made up of polymers of carbohydrates (polysaccharides), polymers of amino acids (proteins) and polyphenolic compounds (lignin). The lignin component is unique to plants, not being present in animal or microbial cell walls. While it might be thought then that fractionation of plant cell wall components may merely involve separation of classes and types of polymers, this is not the whole story. Cell walls show considerable resistance, they do after all have to physically contain the intracellular processes, and so it may be thought that mild treatments are of limited use. However, if the aim is to develop a model for a particular type of cell wall, it should be possible to carry out a series of extraction procedures as mild as possible which will enable us to reconstruct at least a portion of the cell wall complex from a number of "overlapping" analytical procedures. This is the theme for the first chapter in this book. The next chapter is more specific and deals with the analysis of cell wall polymers from olive pulp. It also describes methods for obtaining cell wall material from pulp virtually free of other intracellular compounds.

Autolysis is the name given to the processes of the cell wall where both enzymes and polysaccharides intervene as catalysts, and basically describes cell wall turnover (synthesis and degradation). The chapter on the determination of cell wall autolysis describes the isolation of both autolytically active and inactive cell walls, incubation in buffer and the analysis of autolysis products. There then follows a chapter on the use of enzymes in preparing plant material for electron microscopy for examianation of plant cell walls. This very innovative method has proven very useful in looking at solanaceous material, as described by the author at the recent Fourth International Conference on the Solanaceae in Adelaide.

The methods involved in the determination of cell wall porosity are detailed in the next chapter. This is an important topic since the cell walls act as a molecular and ionic filter and allow passage of some components and restrict movement of others. Methods described include microscopic examination, bulk exclusion techniques and tracer techniques. The following chapter discusses plant cell-substratum adhesives, a fascinating topic, which begins with a short description and gives examples of these adhesives. The chapter concludes with a presentation of the methods involved in studies on adhesives, including assay systems, use of mutants, and physical and chemical methods. No volume devoted to cell wall analysis would be complete without mention of extensin, a structural glycoprotein found in the cell walls of dicotyledonous and monocotyledonous angiosperms, as well as some of gymnosperms. These proteins are bound ionically soon after secretion, but become covalently cross-linked soon after. The chapter devoted to extensin puts forward several methods of analysis of extensin, including biochemical charaterization of the protein, in vivo and in vitro antibody methods of detection and molecular tools for extensin gene studies.

This volume is rounded off by two chapters dealing with possibly the most important plant cell wall components: polymers, lignins and polsaccharides. The chapter on lignin includes a very comprehensive array of methods, from the several methods used to distinguish lignin from tannins and other polyphenols, to colour reactions for lignins, isolation procedures and chemical and physical analysis of lignins. The volume concludes with a chapter on the characterization of plant cell wall-derived oligosaccharides by on-line high-performance anion-exchange chromatography thermospray mass spectrometry.

Acknowledgements. The editors would like to express their gratitude to all contributors for their efforts in keeping the production schedules, and to Dr. Dieter Czeschlik and the staff of Springer-Verlag, especially Ms. Isolde Tegtmeier, and Mr. K.-H. Winter for their cooperation in preparing this and other volumes of the series, *Modern Methods of Plant Analysis*.

Adelaide and Nijmegen/Siena/Amherst, Fall 1995.

H.F. Linskens
J.F. Jackson

Contents

Fractionation of Cell Wall Components
I.M. MORRISON ... 1

1 Introduction ... 1
2 Polysaccharides ... 1
2.1 Precipitation Reactions ... 2
2.1.1 Adjustment of pH ... 2
2.1.2 Precipitation with Organic Solvents ... 2
2.1.3 Precipitation with Inorganic Salts ... 3
2.1.4 Precipitation with Iodine ... 4
2.1.5 Precipitation with Ionic Detergents ... 5
2.2 Chromatography ... 5
2.2.1 Size Chromatography ... 5
2.2.2 Ion Exchange Chromatography ... 7
2.2.3 Affinity Chromatography ... 9
2.3 Electrophoresis ... 10
2.3.1 Moving Boundary Electrophoresis ... 10
2.3.2 Paper Electrophoresis ... 11
2.4 Miscellaneous Polysaccharide Methods ... 11
2.4.1 Cellulose ... 11
2.4.2 Polysaccharide Derivatives ... 11
3 Proteins and Glycoproteins ... 12
3.1 Precipitation Reactions ... 13
3.1.1 Precipitation with Ammonium Sulphate ... 13
3.1.2 Precipitation with Trichloroacetic Acid ... 13
3.2 Chromatography ... 13
3.2.1 Size Chromatography ... 13
3.2.2 Ion Exchange Chromatography ... 13
3.3 Electrophoresis ... 14
3.3.1 Sodium Dodecylsulphate-Polyacrylamide Gel Electrophoresis ... 14
3.3.2 Isoelectric Focusing ... 14
4 Lignins ... 15
References ... 16

Isolation and Analysis of Cell Wall Polymers from Olive Pulp
M.A. COIMBRA, I. DELGADILLO, K.W. WALDRON, and R.R. SELVENDRAN ... 19

1 Introduction 19
2 Isolation of Cell Walls from Olive Pulp 19
2.1 Preparation and Use of Alcohol-Insoluble Residue (AIR) – General Considerations 20
2.2 Preparation of Cell Wall Material (CWM) 21
2.2.1 Material Solubilized During the Preparation of CWM 23
3 Sequential Extraction of Cell Wall Polymers 24
3.1 General Considerations 24
3.2 Sequential Extraction 24
3.2.1 General Comments to the Method 27
4 Fractionation of the Extracted Cell Wall Polysaccharides 29
4.1 General Considerations 29
4.2 Graded Precipitation with Ethanol 30
4.3 Anion-Exchange Chromatography 30
4.3.1 Anion-Exchange Chromatography of Pectic Polysaccharides . . . 32
4.3.2 Anion-Exchange Chromatography of Hemicellulosic Polysaccharides 33
4.4 Fractionation of Acidic Xylans 34
5 Chemical and Spectroscopic Analysis 34
5.1 Neutral Sugars 34
5.2 Uronic Acid 35
5.3 Hydroxyproline Estimation 36
5.4 Methylation Analysis 36
5.4.1 Hakomori Methylation 36
5.4.2 Ciucanu and Kerek Methylation 36
5.4.3 Carboxyl-Reduction of Methylated Polysaccharides 37
5.4.4 Comments on Results of Methylation Analysis 37
5.5 ^{13}C-NMR Studies on Cell Wall Polysaccharides 39
5.6 Fourier Transform Infrared Spectroscopy (FTIR) 39
6 Concluding Remarks 41
References 41

Determination of Cell Wall Autolysis
P.P. Gallego 45

1 Introduction 45
2 Factors Altering Cell Wall Isolation and Autolysis 46
2.1 Plant Material 47
2.2 Buffers 47
2.3 pH 48
2.4 Ionic Strength 48
3 Isolation of Active Cell Walls 49
3.1 Homogenization in Aqueous Medium 49
3.2 Filtration and Washes 50
4 Isolation of Inactive Cell Walls (Controls) 51

4.1 Boiled in Hot Water ... 51
4.2 Boiled in Alcohol ... 51
4.3 Phenol-Acetic Acid-Water ... 52
4.4 Buffer Phenol at pH 7 ... 52
4.5 Low Temperatures ... 53
5 Autolysis Incubations ... 53
5.1 Optimization of the Autolysis Reaction ... 54
5.2 General Procedure ... 54
6 Analysis of Autolysis Products ... 56
6.1 General Methods ... 56
6.2 Ethanol Precipitation ... 56
6.3 Gel Permeation Chromatography ... 57
6.4 Ion Exchange Chromatography ... 57
7 Conclusions ... 57
References ... 58

Cell Wall Porosity and Its Determination
S.M. Read and A. Bacic ... 63

1 Introduction ... 63
1.1 Biological Significance of Wall Porosity ... 63
1.2 Assumptions and Definitions ... 64
2 Microscopic Visualization of Wall Pores ... 66
3 Bulk Exclusion Techniques ... 67
3.1 Solute Exclusion ... 67
3.2 Long-Term-Exclusion of PEG (Polyethylenglycol) ... 69
3.3 Gel Filtration ... 70
4 Tracer Techniques: Uptake of Molecules or Particles ... 71
4.1 Small Molecules ... 71
4.2 Macromolecules ... 72
4.3 Particles ... 75
5 Conclusions ... 75
5.1 Integration of Results from Different Methods ... 75
5.2 Pore Structure ... 76
5.3 Variations in Wall Porosity ... 77
5.4 Future Developments ... 78
References ... 78

Analysis of Chitin Biosynthesis
M. Fèvre, L. Gay, and H. Chanzy ... 81

1 Introduction ... 81
2 Structure of Chitin ... 81
3 The Enzymatic Synthesis of Chitin ... 82

3.1 Assay of Chitin Synthase ... 83
3.2 Inhibition of Enzymatic Activity ... 83
4 Preparation of Chitin Synthase ... 84
5 Product Characterization of Chitin Synthase ... 85
5.1 Chemical Identification ... 85
5.2 Biophysical Identification of Chitin ... 86
5.2.1 Fourier Transform Infrared Spectroscopy ... 86
5.2.2 X-Ray Diffraction Analysis ... 86
5.2.3 Electron Microscopy and Electron Diffraction Analysis ... 88
6 Polymerization and Crystallization ... 90
7 Fungal Chitin Synthase Genes ... 90
8 Morphogenetic Roles of Chitin Synthases ... 91
9 Conclusions ... 92
References ... 92

Analysis of Plant-Substratum Adhesives
V. Vreeland and L. Epstein ... 95

1 Introduction ... 95
1.1 Principles of Adhesion ... 95
1.2 Cell-Substratum Adhesion ... 96
1.2.1 Higher Plant-Substratum Adhesion ... 98
1.2.2 Adhesion of Fungal Phytopathogens to a Plant Substratum ... 98
1.2.3 Algal-Substratum Adhesion ... 99
2 Case Studies of Cell-Substratum Adhesion ... 100
2.1 Substrate Adhesion by the *Fucus* Zygote ... 100
2.2 Adhesion of Conidiospores of the Plant Pathogenic Fungus, *Nectria haematococca* ... 103
3 Methods for the Analysis of Cell-Substratum Adhesives ... 104
3.1 Adhesion Systems and Assays ... 104
3.2 Identification of Adhesive Components ... 105
3.2.1 Isolation of Adhesion Mutants ... 105
3.2.2 Correlation of Temporal/Spatial Development with Adhesion ... 106
3.2.3 Experimental Perturbation of Adhesion ... 107
3.3 Analysis of Adhesive Components ... 108
3.3.1 Extraction and Purification ... 108
3.3.2 Assays of Isolated Compounds for Adhesive Activity ... 109
4 Conclusions ... 110
References ... 111

Biochemical, Immunological and Molecular Analyses of Extensin
E.E. Hood ... 117

1 Introduction 117
2 Biochemical Characterization of Extensin 117
2.1 Preparation of Walls 117
2.2 Protein Purification 118
2.3 Biochemical Characterization 119
3 Immunological Detection of Extensin 120
3.1 In Vivo Localization of Extensin 121
3.2 Western Blots and Tissue Prints 121
4 Molecular Characterization of Extensin 122
4.1 Isolation of Genes Encoding Extensin 123
4.2 Extensin Expression Studies 125
4.3 Extensin Promoter Fusions with Reporter Genes 125
4.4 Generation of Extensin Mutants 126
5 Summary 126
References 126

Analysis of Pectin Structure by HPAEC-PAD
A.T. Hotchkiss, Jr., K. El-Bahtimy, and M.L. Fishman 129

1 Introduction 129
2 Pectin HPLC Separations 130
2.1 GPC 130
2.2 Ion-Exchange and Ion-Pair RP 130
2.3 HPAEC-PAD 131
3 Pectin Analysis in Ripening Peach Fruit 134
3.1 Melting- and Nonmelting-Flesh Peaches 134
3.2 HPAEC-PAD System 135
3.3 Sample Preparation 136
3.4 HPAEC-PAD of Redskin and Suncling Peach Pectin 136
4 Conclusions and Future Directions 142
References 142

Characterization of Oligosaccharides Derived from Plant Cell Wall Polysaccharides by On-Line High-Performance Anion-Exchange Chromatography Thermospray Mass Spectrometry
W.M.A. Niessen, H.A. Schols, R.A.M. van der Hoeven, A.G.J. Voragen, and J. van der Greef 147

1 Introduction 147
2 Methods 148
2.1 Apparatus 148
2.2 Isolation of Oligosaccharides from Plant Cell Wall Material 150
3 HPAEC in Oligosaccharide Analysis 150

4 General Experimental Considerations Related to HPAEC-MS 152
4.1 Desalting by AMMS 152
4.2 Ionization of Oligosaccharides 153
4.3 Data Interpretation 155
5 Application of HPAEC-MS in Oligosaccharide Characterization 156
6 Conclusions and Perspectives 160
References 162

Analysis of Pectin Methyl Esterases
M. Bordenave 165

1 Introduction 165
2 Estimation and Detection of PME Activities 165
2.1 Detection 165
2.2 Estimation 167
3 PME Localization 167
4 PME Extraction and Purification 168
5 PME Properties 169
5.1 Physicochemical Properties 169
5.2 Enzymic Properties 171
5.2.1 Action Pattern 171
5.2.2 Influence of pH on PME Activity 172
5.2.3 Influence of Cations 174
5.3 Structures 174
6 Roles of PMEs 175
6.1 Plant Pathogens 175
6.2 Fruit Maturation 176
6.3 Cell Elongation 176
7 Conclusions and Perspectives 177
References 177

Probing the Subunit Composition and Topology of Plasma Membrane-Bound (1,3)-β-Glucan (Callose) Synthases
B.P. Wasserman, Xiaoqun Qi, L.M. Barone, and Ayong Wu 181

1 Introduction 181
2 Membrane Isolation, Enzyme Assay, and Solubilization 183
2.1 Isolation of Crude Membrane Fractions 183
2.2 Plasma Membrane Isolation 185
2.3 Membranes of Defined Sidedness: Inside Out and Right-Side Out Vesicles 185
2.4 Callose Synthase Assay 185
2.5 Callose Synthase Solubilization 186

3 Callose Synthase Topology, Purification, and Subunit Composition . . . 187
3.1 Polypeptide Depletion . . . 187
3.2 Vectorial Proteolysis . . . 188
3.3 Callose Synthase Purification . . . 189
3.3.1 Glycerol Gradient Centrifugation . . . 189
3.3.2 Product Entrapment . . . 191
4 Biochemical Characterization of Integral Plant Plasma Membrane Proteins . . . 191
4.1 Characterization of Disulfide-Linked Aggregate Formation . . . 191
4.2 Recovery of Hydrophobic Plant-Derived Membrane Proteins for Sequencing and Antibody Production . . . 192
4.3 Antibody Characterization . . . 193
4.3.1 Immunoblotting . . . 194
4.3.2 Affinity Purification of Anti-PMIP27 . . . 194
5 Summary . . . 195
References . . . 196

Subject Index . . . 199

List of Contributors

Bacic, Antony, Plant Cell Biology Research Centre, School of Botany, University of Melbourne, Parkville, Victoria 3052, Australia

Barone, Lucille M., Department of Food Science, New Jersey Agricultural Experiment Station, Cook College, Rutgers University, New Brunswick, New Jersey 08903-0231, USA

Bordenave, Marianne, Laboratoire d'Enzymologie en Milieu structuré, Département de Biologie Supramoléculaire et Cellulaire, Institut Jacques Monod, 2, place Jussieu, 75251 Paris Cedex 05, France

Chanzy, Henri, Centre de Recherches sur les Macromolécules Végétales - CERMAV CNRS BP 53 X, 38041 Grenoble Cedex 9, France

Coimbra, Manuel A., Department of Chemistry, University of Aveiro, 3800 Aveiro, Portugal

Delgadillo, Ivonne, Department of Chemistry, University of Aveiro, 3800 Aveiro, Portugal

El-Bahtimy Karimah, US Department of Agriculture, Agricultural Research Service, Eastern Regional Research Center, Philadelphia, Pennsylvania 19118, USA

Epstein, Lynn, Department of Environmental Science, Policy and Management, 147 Hilgard Hall, University of California, Berkeley, California 94720, USA

Fèvre, Michel, Laboratoire de Biologie Cellulaire Fongique, Centre de Génétique Moléculaire et Cellulaire - UMR CNRS 106, Université Lyon I - 43 bd du 11 Novembre 1918, 69622 Villeurbanne Cedex, France

Fishman, Marshall L., US Department of Agriculture, Agricultural Research Service, Eastern Regional Research Center, Philadelphia, Pennsylvania 19118, USA

Gallego, Pedro P., Universidad de Vigo, Lab. Fisiologia Vegetal. Fac. Ciencias, Apdo. 874, Vigo. Pontevedra, 36200 Spain

Gay, Lucien, Laboratoire de Biologie Cellulaire Fongique, Centre de Génétique Moléculaire et Cellulaire - UMR CNRS 106, Université Lyon I - 43 bd du 11 Novembre 1918, 69622 Villeurbanne Cedex, France

Hood, Elizabeth E., Department of Biology, Utah State University, Logan, Utah 84322-5305, USA

Hotchkiss, Arland T., US Department of Agriculture, Agricultural Research Service, Eastern Regional Research Center, Philadelphia, Pennsylvania 19118, USA

MORRISON, IAN M., Scottish Crop Research Institute, Invergowrie, Dundee DD2 5DA, UK

NIESSEN, W.M.A., De Wetstraat 8, 2332 XT Leiden, The Netherlands

QI, XIAOQUN, Department of Food Science, New Jersey Agricultural Experiment Station, Cook College, Rutgers University, New Brunswick, New Jersey 08903-0231, USA

READ, STEPHEN M., Plant Cell Biology Research Centre, School of Botany, University of Melbourne, Parkville, Victoria 3052, Australia

SCHOLS, H.A., Wageningen Agricultural University, Department of Food Science, Bomenweg 2, 6703 HD Wageningen, The Netherlands

SELVENDRAN, ROBERT R., Department of Food Molecular Biochemistry, AFRC Institute of Food Research, Norwich Research Park, Colney, Norwich NR4 7UA, UK

VAN DER GREEF, J., Division of Analytical Chemistry, Leiden Amsterdam Center for Drug Research, P.O. Box 9502, 2300 RA Leiden, The Netherlands

VAN DER HOEVEN, R.A.M., Division of Analytical Chemistry, Leiden Amsterdam Center for Drug Research, P.O. Box 9502, 2300 RA Leiden, The Netherlands

VORAGEN, A.G.J., Wageningen Agricultural University, Department of Food Science, Bomenweg 2, 6703 HD Wageningen, The Netherlands

VREELAND, VALERIE, Department of Plant Biology, 111 Koshland Hall, University of California, Berkeley, California 94720, USA

WALDRON, KEITH W., Department of Food Molecular Biochemistry, AFRC Institute of Food Research, Norwich Research Park, Colney, Norwich NR4 7UA, UK

WASSERMAN, BRUCE P., Department of Food Science, New Jersey Agricultural Experiment Station, Cook College, Rutgers University, New Brunswick, New Jersey 08903-0231, USA

WU, AYONG, Department of Food Science, New Jersey Agricultural Experiment Station, Cook College, Rutgers University, New Brunswick, New Jersey 08903-0231, USA

Fractionation of Cell Wall Components

Ian M. Morrison

1 Introduction

The major constituents of plant cell walls are carbohydrates, but other classes of constituents are present. Proteins are present in varying concentrations depending, in part, on the plant material being examined and the type and maturity of the tissue being investigated. Some of the proteins are wall-bound enzymes: the rest are structural elements. A unique feature of plant cell walls, compared to animal and microbial walls, is the presence of the polyphenolic material, lignin. Although lignin is usually defined as singular, there is increasing evidence that different proportions of structural elements are present and probably different linkages are prevalent in different tissues. With improved analytical techniques it may turn out that these constituents should be classed in the plural.

Methods for the fractionation or separation of classes of constituents isolated from plant cell walls depend on the subsequent use of the materials or information. Fractionation should not purely be considered as the separation of classes and types of polymers for subsequent utilization or analysis on a qualitative basis: the term fractionation is equally applicable to qualitative and analytical separations where the constituents are not isolated and only identified, for example, by spectroscopic techniques.

A criticism of fractionation techniques, particularly for the analysis of cell wall components, is that they only determine a portion of a much greater complex. Since one of the functions of cell walls is to contain intracellular processes, the cell wall must show considerable resistance. It is not surprising, therefore, that significant proportions of plant cell walls are not removed by mild treatments such as extraction with water. While some constituents are removed by mild aqueous treatments such as the use of chelating agents, the greatest proportion requires much stronger treatments which have inevitably broken inter-polymeric covalent bonds. However, if the aim is to develop a model for a particular type of cell wall, it should be possible to use a range of extraction methods which are as mild as possible, but which also provide overlapping sequences to construct at least a fragment of the original macromolecular complex from which they were derived.

2 Polysaccharides

The polysaccharides are, by weight, the major components of plant cell walls. Cellulose is an ubiquitous constituent of all land plants and the primary structure,

Modern Methods of Plant Analysis, Vol. 17
Plant Cell Wall Analysis
Edited by H.-F. Linskens and J.F. Jackson

being a homopolysaccharide containing β-(1,4)-linked D-glucopyranosyl residues, does not vary. The other polysaccharides are far more diverse. The current view is to group them all together under the general umbrella of non-cellulosic polysaccharides (NCPs). This includes the acidic polysaccharides (pectins) which can contain 20 to 90% acidic sugar (galacturonic acid residues) and the neutral polysaccharides (hemicelluloses), some of which can have up to 10% acidic sugar. The constituents in the latter are very diverse and this diversity in structure allows fractionation of apparently closely related species. Their nomenclature depends on the major sugar residue or the sugar residues of the main chain. For example, the arabinoxylans are a family of polymers in which the D-xylopyranosyl residues constitute the main chain with the L-arabinofuranosyl residues being present in side chains, while the glucomanans are a family of polymers in which both D-mannopyranosyl and D-glucopyranosyl residues are present in the main chain but the mannose content is greater than the glucose content. The latter type can also have side chains as found in the galactoglucomannans. Other major families are the xyloglucans, the arabinogalactans and the mixed linked β-glucans.

2.1 Precipitation Reactions

2.1.1 Adjustment of pH

One of the earliest, and still very widely used, fractionation procedures is that used for the preliminary fractionation of the hemicelluloses. Some authors define hemicelluloses as the non-pectic polysaccharides of the plant cell wall which are extracted with alkali. Once the range of polysaccharides has been extracted with alkali, adjustment of the pH to 4 with acid brings about the precipitation of the more linear of the range of xylans (O'Dwyer 1926). Acetic acid is the acid traditionally used since it is a weak acid and the pH is not inadvertently reduced to a value where hydrolysis of acid-labile groups could occur. However, since the realization that acetate esters are covalently bound to cell wall polysaccharides, some authors have used inorganic acids, such as hydrochloric acid to allow the determination of acetate groups released by the alkaline extraction. No method is ideal; a subsequent precipitation with ethanol will have a lower salt contamination if acetic acid is used. In the O'Dwyer classification, the material precipitated on acidification is termed hemicellulose A.

2.1.2 Precipitation with Organic Solvents

Polysaccharides are insoluble in organic solvents so they can be precipitated from a neutral aqueous solution. Precipitation can be total at a concentration of 80–90% ethanol or methanol. After an alkaline extraction, some polymeric material may remain soluble in 90% ethanol but this is usually assumed to be partially degraded. Naturally, the organic solvent must be miscible with water, and ethanol or methanol are normally used. Acetone and propan-2-ol have also been used. In the O'Dwyer classification, this fraction is designated hemicellulose B.

The concentration of ethanol required to precipitate a polysaccharide from aqueous solution varies considerably depending on the polymer. Therefore, fractional precipitation can be achieved by gradually increasing the alcohol concentration. This is usually done in increments of 10%: the precipitated material at each stage is centrifuged off and the supernatant is retreated with more ethanol to reach the next increment. Each fraction must be analysed, usually be acidic hydrolysis, to determine which fractions have a similar composition. In a typical experiment, Brillouet et al. (1982) isolated a heteroxylan from an extract from wheat bran. The fractions which were precipitated at concentrations between 60 and 90% ethanol had very similar compositions and accounted for 70% of the hemicellulose fraction. The fractions precipitated at lower ethanol concentrations had increasing arabinose: xylose ratios but all were considerably lower than those in the main fraction.

2.1.3 Precipitation with Inorganic Salts

2.1.3.1 Precipitation with Cu(II) Ions

There are two fractionation methods using Cu(II) ions which have been used extensively. The first uses copper acetate and was proposed by Erskine and Jones (1956) for a mixture containing acidic and neutral polysaccharides. A 7% aqueous solution of copper acetate is added to an aqueous solution of the mixture until no further precipitate of the Cu(II) salt of the acidic polysaccharide is formed. After centrifuging off the precipitate, the neutral polysaccharide, which also forms a copper complex, is precipitated by adding 1 vol of ethanol. The polysaccharide-copper complexes are decomposed with 1% hydrochloric acid in ethanol, washed with ethanol and dried.

The second method is for the fractionation of water-insoluble but alkali-soluble polysaccharides and makes use of Fehling's solution (Adams 1957). The polysaccharide mixture is suspended in water (1 g:40 ml) and an equal volume of 2 M sodium hydroxide is added. Fehling's solution is added dropwise until precipitation is complete. After centrifuging off the precipitate, further polysaccharide is recovered by adding an equal volume of ethanol. The polysaccharide-copper complexes are decomposed by treatment with 1% HCl in ethanol and the polysaccharide is dried by solvent exchange.

Procedures for the fractionation of polysaccharides via copper complexes have been detailed by Jones and Stoodley (1965).

2.1.3.2 Precipitation with Al(III) Ions

For the preparation of pectin, particularly on an industrial scale, it can be precipitated as a complex with aluminium hydroxide (Joslyn and de Luca 1957). The pH of a solution of the extracted pectin is adjusted to < 3.5 and either aluminium chloride or sulphate is added as a concentrated solution. The pH is then raised to 4.2–4.5 which converts the aluminium salt to the positively charged hydroxide and which precipitates with the negatively charged pectin. The A1(III) ion is then removed by washing with acidic ethanol; the pectin is then dried.

2.1.3.3 *Precipitation with Ba(II) Ions*

Mannans, glucomannans and galactomannans can be precipitated as a Ba(II) complex if the 2,3 *cis* diol configuration is free on the D-mannose residues. Most, if not all, plant mannans are 1,4-linked and the D-glucose residue of the glucomannans is in the main chain (i.e. also 1,4-linked), while the D-galactose residues are linked at C-6 of the mannopyranose rings of galactomannans. The above requirement is therefore met. The polysaccharide mixture is dissolved in water or 0.1 M NaOH, depending on the solubility, and a 0.1 M solution of barium hydroxide is added dropwise with stirring until no further precipitate appears. After centrifuging off and washing with barium hydroxide solution, the complex is decomposed with acetic acid and the purified polysaccharide recovered (Meier 1965; Seth et al. 1984).

2.1.3.4 *Precipitation with Borate Ion*

Some success has been achieved by precipitation with borate, which also makes use of its ability to complex with the 2,3 *cis* diol configuration of the D-mannopyranose residues. However, more success has been achieved with borate in an affinity chromatographic role; this ion will be discussed in more detail in Section 2.2.3.2.

2.1.4 Precipitation with Iodine

Water-soluble polysaccharides can be fractionated into linear and branched species by precipitation with iodine (Gaillard 1961).

The hemicellulose B fraction (200 mg) from maize cobs is dissolved in aqueous calcium chloride solution (20 ml: specific gravity 1.3). Addition of an iodine solution (2.0 ml: $30\,g\,l^{-1}$ iodine and $40\,g\,l^{-1}$ potassium iodide) results in the precipitation of a dark blue precipitate which is separated by centrifugation at $20\,000 \times g$. The clear supernatant solution is pipetted off and the excess iodine is reduced to iodide ion with the thiosulphate ion. The polysaccharide in this soluble fraction is precipitated with ethanol (4 vol). The blue iodine/polysaccharide complex is decomposed by suspension in water and addition of a small amount of sodium thiosulphate; the polysaccharide is recovered by precipitation in ethanol (4 vol).

In the above example, the polymer precipitated as the iodine complex was essentially a linear polysaccharide which contained D-xylose and L-arabinose residues in the ratio 8.4 : 1, while the polymer remaining in solution was a branched polysaccharide containing D-xylose, L-arabinose and D-galactose residues in the ratio 7.9 : 2.7 : 1 as well as some uronic acid residues. The method has been used to compare the linear and branched polysaccharides in the hemicellulose B fraction of three members of the Gramineae family and three members of the Leguminosae family (Gaillard 1965). It has also been used to determine the changes in linear and branched polysaccharides with increasing maturity from several grasses (Morrison 1980).

2.1.5 Precipitation with Ionic Detergents

The use of ionic detergents to precipitate acidic polysaccharides was first proposed by Scott (1960) and a specific procedure using cetyltrimethylammonium bromide (CTAB) has been reviewed (Scott 1965). An aqueous solution of the polysaccharide mixture (ca. 1 g in 800 ml) is treated with 0.1 M CTAB until no further precipitate appears. Any neutral polysaccharide remaining in solution can be recovered by precipitation in ethanol.

A variation of the method was proposed by Aspinall et al. (1966). They converted the detergent to its weak base, cetyltrimethylammonium hydroxide, by passing the solution of CTAB through a column of Amberlite IRA400(OH). The weak base causes less degradation than CTAB itself. The procedure was used to remove contaminating acidic polysaccharides and to leave a pure "mannan" in solution.

2.2 Chromatography

2.2.1 Size Chromatography

One of the most frequently and universally used fractionation procedures is that based on chromatography by size separation. The technique has been available for nearly 40 years and has been considered under a number of titles in that time. The early papers used the term gel filtration, but more modern procedures use such terms as gel permeation, molecular sieve or size exclusion chromatography. All rely on the use of cross-linked matrices. The size of the pores in these materials is rigorously controlled during manufacture, but there is inevitably a range of sizes in any such materials. It is this range of sizes that allows a separation. Molecules which are too large to enter any of the pores are simply washed through and are eluted from the column in the void volume (V_o). Small molecules are able to enter all the pores and are retained to the greatest extent. They are eluted at the total volume of the column (V_t). Molecules of intermediate size are retained to intermediate extents: the actual point of elution, as a proportion of the included volume ($V_i = V_t - V_o$), can be used to calculate an approximate molecular size.

2.2.1.1 Low Pressure Systems

The following materials are suitable for low pressure applications. They form soft gels which are easily deformed under pressure so strict conditions must be applied. In general, for a given type of matrix, the lower the proportion of cross-links, the lower the permitted pressure. The matrices are manufactured from a variety of polymers: the cross-links are introduced during the manufacture of the matrix and use a wide range of bifunctional molecules. For example, one of the earliest, and still used, matrices is Sephadex which has a polysaccharide base of dextran, an α 1,6-linked D-glucan of bacterial origin, which is cross-linked with epichlorohydrin. The different Sephadexes (G-10 to G-200) have separation ranges from 0–700 up to 5–250 × 10^3 Da respectively. The other most frequently used materials in the same fractionation ranges are the Bio-Gels which are prepared by copolymerization of acrylamide and *N,N'*-methylene-bis-acrylamide. These materials (Bio-Gel P-2 to P-100) have nominal separation ranges of 100–1800 to 5–100 × 10^3 Da respectively. Both ranges require an aqueous-based solvent but urea, detergents, chaotropic reagents and low concentrations of ethanol are compatible. They can be used over a pH range of ca. 2 to 10.

For the fractionation of materials with higher molecular sizes, other materials are available. Sephacryl is a copolymer of allyl dextran and *N,N'*-methylene-bis-acrylamide and products can fractionate in the range 1–100 000 × 10^3 Da, whereas the Sepharoses (beaded agarose) cover the range 10–20 000 × 10^3 Da. The Bio-Gel A series are similarly based on beaded agarose and nominally cover the range 10–150 000 × 10^3 Da.

While the procedure is fully documented and has been applied to innumerable separations, a number of considerations must still be made. The molecular size attributed to each fraction will depend on the actual shape of the molecule. A spherical molecule of specific molecular weight is able to enter a pore more easily than a linear molecule of the same molecular weight. For example, the separation range for Sephadex G-150 is 5–300 × 10^3 Da for globular molecules and only 1–150 × 10^3 Da for linear dextrans. Hence, molecular weights determined by size exclusion chromatography are only realistic if the precise shape of the polysaccharide is known. Size exclusion chromatography is less precise for plant cell wall polysaccharides than for proteins since cell wall polysaccharides are usually polydisperse and, while separation can be achieved, the elution patterns from a plant extract will usually overlap. It is only from special cases such as suspension culture products or when fragments are released from a native polysaccharide, especially by enzymic hydrolysis, that they are sufficiently homogeneous to give a sharp chromatogram. For example, the xyloglucan-derived oligosaccharides XXFG (XG9) and XXXG (XG7) can be separated on Bio-Gel P-2 (Hayashi and Maclachlan 1984). A final consideration applies specifically to Sephadex G-25 and those with lower G numbers. If the polysaccharide carries phenolic constituents, such as the phenolic acids which are present on the xylans of the Gramineae, anomalous size values may result. Values greater than the included volume of the column have been reported and are considered to be due to an interaction between the phenolic component and the epichlorohydrin residue which was used for cross-linking.

2.2.1.2 *High Performance (Pressure) Systems*

For high performance (pressure) systems, soft gels such as Sephadex cannot be used. Several types are available and the following are merely given as examples with no recommendation being made regarding preference. The Bio-Sil SEC range has a silica matrix and can operate over the range 5–1000 × 10^3 Da, while the Bio-Gel SEC range is quoted as "polymer-based" but has a separation range of 1–200 000 × 10^3 Da. Both types are marketed in analytical and preparative versions with maximum capacities of 1 and 100 mg respectively. The polymer-based matrix also offers a wider pH range of 2–12, while the silica-based matrix is limited to 2–7. Another range is BioSep-SEC, also silica-based but with a separation range of 1–2000 × 10^3 Da.

The benefits of HPLC systems can be seen in the separation of the xyloglucan-derived oligosaccharides XXXG (XG7) and XXFG (XG9) on Amino-Spheri-5 and eluting with 60% acetonitrile in water at room temperature (Fig. 1; McDougall and Fry 1991).

2.2.2 Ion Exchange Chromatography

Ion exchange chromatography is generally a more efficient methods for fractionation of plant cell wall polysaccharides and their degradation products. The pectins contain high concentrations (20–100%) of galacturonic acid residues, though the uronic acid groups of the less degraded fractions are masked by the

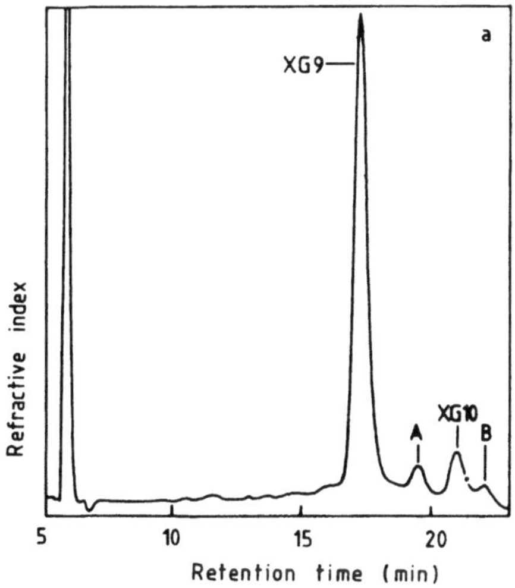

Fig. 1. HPLC separation of xyloglucan oligosaccharides XXXG (*XG7*) and XXFG (*XG9*) on Amino-Spheri-5. (McDougall and Fry 1991)

presence of methyl esters. Nonetheless, they and their enzymic hydrolysis products are readily separable by anion exchange chromatography. Some of the other non-cellulosic polysaccharides, such as the hemicellulosic arabinoxylans, also carry uronic acid residues, though at a far lower concentration (<5%), but they too are sufficiently charged to bind to the same exchangers. In addition, when a mixture contains both neutral and acidic polysaccharides, it is readily fractionated by this procedure since the neutral polymers are eluted with water or buffers of low ionic strength, while the acidic polymers require buffers of high ionic strength. Cation exchange chromatography is less useful in polysaccharide fractionation since few such polymers are basic. The exceptions are the chitins and chitosans which are found in the walls of some fungi.

2.2.2.1 Anion Exchange Chromatography

The most frequently used anion exchange resins for the separation of polysaccharides use the diethylaminoethyl (DEAE) group as the reactive site, but diethyl(2-hydroxypropyl)aminoethyl (QAE) groups are also used. The matrices used are varied, ranging from dextrans (e.g. Sephadex) and agarose (e.g. Sepharose and Bio-Gel A) to cellulose. A further consideration is the anion form in which the resin is generated. Due to its high reactivity, the hydroxyl ion is rarely used except to remove material which has bound so strongly to the resin that it cannot be removed even by high salt concentrations. The resins are most frequently generated in the phosphate, acetate or chloride forms but the borate form has specific properties. There is some debate whether the borate form of an anion exchange resin really acts as an ion exchanger or as an affinity ligand. Our view is the latter and it is discussed in Section 2.2.3.2. There are numerous references but a specific example showing the effect of the anion is shown in Table 1 (Stevens and Selvendran 1984).

Table 1. Effect of the ionic form on the separation of alkali-soluble hemicelluloses from cabbage leaves by anion exchange chromatography. (Stevens and Selvendran 1984)

Fraction	Sugar composition (mg g^{-1})						
	Rha/Fuc	Ara	Xyl	Man	Gal	Glc	UA
A	41	95	165	63	120	241	74
B	23	116	315	28	69	62	84
A1	56	54	178	100	128	362	57
A2	54	179	205	33	125	297	103
A3	4	30	92	14	13	39	22
A4	2	15	42	20	nd	27	nd
A5	6	40	38	3	21	18	18

Fractions A and B were separated on DEAE-Sephacel (Cl^- form), while A1 to A5 were separated by rechromatographing sample A on DEAE-Sephacel (Ac^- form).

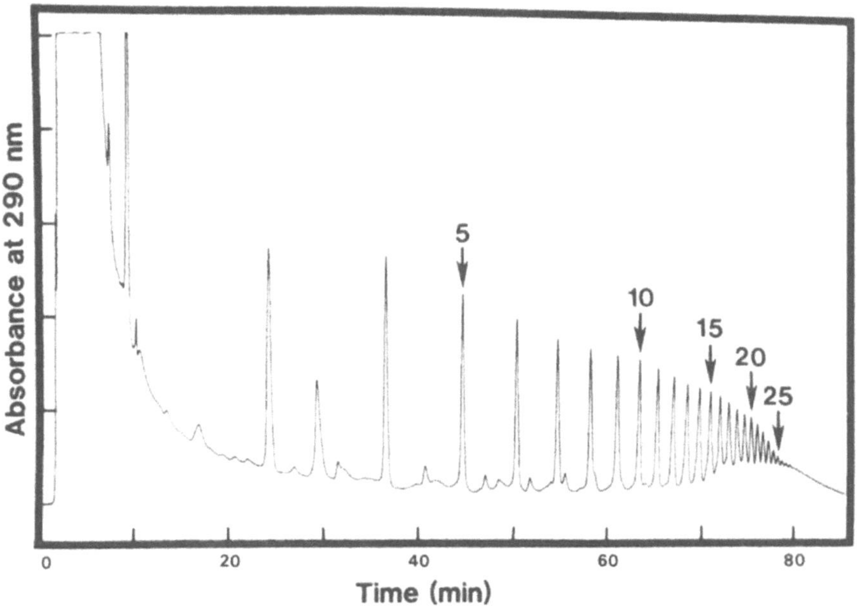

Fig. 2. Anion-exchange HPLC of 2-aminopyridinyl oligogalacturonides from hydrolysis of pectate on Spherogel TSK DEAE-2SW. The peaks represent 10 to 0.2 nmol oligogalacturonide. (Maness and Mort 1989)

2.2.2.2 Cation Exchange Chromatography

Cation exchange resins mainly use carboxymethyl (CM) and sulphopropyl (SP) groups for the reactive sites and the matrices are the same as those used for anion exchange chromatography.

The most practical and quantitative applications of ion-exchange chromatography to the fractionation of plant polysaccharides have been used in low pressure systems. High performance systems have been used but are more effective for the separation of molecules with lower molecular weights. An example of the anion-exchange HPLC of oligogalacturonides is shown in Fig. 2 (Maness and Mort 1989). The range of oligogalacturonic acids was isolated from an enzymic digest and labelled at their reducing end with 2-aminopyridine to improve the separation and assist the detection. The analytical column was Spherogel TSK DEAE-2SW. Separation of oligomers from 3 to 25 residues was possible.

2.2.3 Affinity Chromatography

2.2.3.1 Affinity Chromatography Using Lectins

Far fewer applications of affinity chromatography techniques for the fractionation of plant cell wall polysaccharides than for animal and microbial polysaccharides have been reported. However, over 20 years ago, Koleva and Achtardjieff (1973) reported the purification of a glucomannan from *Arum orientale* by affinity chromatography on a column of concanavalin A which had been cross-linked to Bio-gel P10 via glutaraldehyde. The column was eluted with methyl α-mannopyranoside. Since then, many other lectins, which have been immobilized by cross-linking to a wide variety of inert matrices, have become commercially available. It is also a relatively simple procedure to cross-link any novel lectin with a suitable matrix to create one's own affinity material (Dean et al. 1985). The procedure for fractionation is fairly standard once the specificity of the lectin has been established.

2.2.3.2 Affinity Chromatography with Borate Derivatives

Anion exchange resins, such as DEAE-Sephedex, can be converted to the borate form with 5 mM disodium tetraborate at pH 9.0–9.5 and can be used as affinity columns for borate-binding polymers such as mannans. Polymers which are unable to bind to borate are eluted with water or very low concentrations of the disodium tetraborate, while the bound species are eluted with high concentrations (ca. 200 mM). If necessary, but less desirable, 0.1 M NaOH can be used. More specific affinity matrices, such as *m*-aminophenylboronic acid-activated agarose (Vlassara et al. 1981), are available but there are no reported applications to mannose-containing plant cell wall polymers. They may be more applicable to glycoproteins.

2.2.3.3 Affinity Chromatography with Insoluble Polysaccharides

Another form of affinity chromatography makes use of the similarity in structure between different polysaccharides. The attempted fractionation of a "pectin" and xylan mixture from soybean hulls on cellulose powder gave a purified pectic fraction, but the xylan could not be recovered even after elution with dilute alkali (Aspinall et al. 1966). The linear xylan was evidently too similar to the cellulose. Later work with a polysaccharide which had a higher degree of substitution did allow the release with alkali (Aspinall et al. 1969).

By contrast, the nona- and heptasaccharide fragments from xyloglucan fragments do not bind to cellulose under physiological conditions but do so in aqueous organic solvents (Valent and Albersheim 1974). They are readily recovered by eluting with dilute alkali. The intact xyloglucans do bind to cellulose (Hayashi et al. 1987) at a ratio of 1:20 and can also be released with alkali.

2.3 Electrophoresis

2.3.1 Moving Boundary Electrophoresis

This procedure is purely analytical and seems to have fallen out of favour, but it does give another accurate assessment of the homogeneity (or lack of it) of samples. For example, pectin preparations from various sources and with different galacturonic acid contents have been analysed in a sodium tetraborate buffer. The results showed that the preparations with a high galacturonic acid content were homogeneous, but those with a low acid sugar content contained at least one other component (Zitko and Bishop 1965).

2.3.2 Paper Electrophoresis

Electrophoresis on glass fibre paper is a relatively simple technique which allows separations based on another set of criteria (Jarvis et al. 1977). The paper is trimethylsilylated to minimize the electroendosmotic effect and a surfactant has to be added to the buffer to allow wetting. It is most effective with charged polymers: the rate of de-esterification of potato pectin by the action of sodium carbonate can be followed by this procedure. In the preparative mode, strips are cut from the electrophoretogram and the appropriate bands are located by suitable stains. The method can also be used for neutral polysaccharides in which a charge can be induced, particularly when a borate buffer is used to generate a polysaccharide-borate complex.

Cellulosic paper can also be used. Feruloyl esters of cell wall fragments, which are extremely alkali labile, have been separated without hydrolysis (Fry 1983).

2.4 Miscellaneous Polysaccharide Methods

2.4.1 Cellulose

While the primary structure of the celluloses from all determined sources is the same, there are differences in the average molecular size of the polymers from different sources. Until recently, most methods to "dissolve" cellulose have caused significant depolymerization/degradation, but cellulosic fibres can be dissolved in dimethylacetamide/lithium chloride and fractionated by gel permeation chromatography on a series of Ultrastyrogel columns using both a viscometer and a refractive index detector (Timpa 1991). The molecular weight distribution pattern (Fig. 3) clearly shows the difference in molecular size between primary and secondary cell wall celluloses.

2.4.2 Polysaccharide Derivatives

While not strictly classed as fractionation, polysaccharides can be converted to their peracetyl derivative by reaction with acetic anhydride in pyridine (Gardner and Chang 1974) or their permethyl derivative by reaction with the dimsyl anion (sodium hydride in dimethyl sulphoxide) and methyl iodide (Hakomori 1964). The change from materials soluble in aqueous solvents to those soluble in organic solvents adds another dimension to fractionation techniques. The derivatized

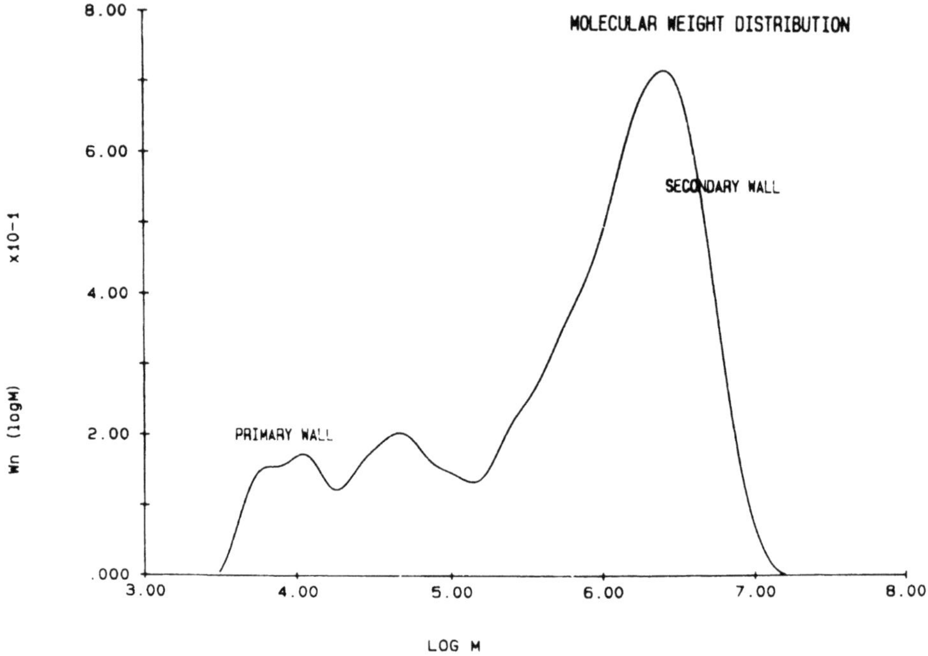

Fig. 3. Molecular weight distribution of cotton fibre samples. (Timpa 1991)

polymers can be fractionally precipitated from chloroform/methanol solutions into petroleum ether and they can also be separated by size chromatography on a variety of non-aqueous systems. Derivatization does have certain drawbacks. Firstly, the derivatization procedure must not cause any degradation. Labile constituents could be removed if acidic or alkaline conditions have to be used. Secondly, if the homogeneous polymer has to be recovered, methylation cannot be used.

3 Proteins and Glycoproteins

Confirmation of the presence of (glyco)proteins in plant cell walls is relatively new and the range and importance are increasing steadily. There are probably two types of (glyco)protein, one having a distinct structural role and the other consisting of proteins which are present in the wall but which probably only have an enzymic function. The former group is recognized as having three distinct subgroups, the hydroxyproline-rich glycoproteins (HRGPs or extensins), the proline-rich proteins (PRPs) and glycine-rich proteins (GRPs) (Keller 1993; Josè and Puigdomènech 1993). A further group ("expansins") has recently been proposed. Most of the cell-wall-associated enzymes are reported to be involved in lignification and include the peroxidases (Welinder et al. 1993) and certain oxidases.

The major difficulty with the proteins is not in fractionation or characterization but in their extraction. The principles of most of the fractionation procedures have already been described in Section 2. Only specific examples will be given here.

3.1 Precipitation Reactions

3.1.1 Precipitation with Ammonium Sulphate

The (glyco)protein preparation is dissolved in the minimum quantity of cold (15–20 °C) water and solid ammonium sulphate is added with stirring. Ammonium sulphate has a solubility of ca. 700 $g\,l^{-1}$ at that temperature. Precipitation is usually initiated at ca. 20% saturation (140 $g\,l^{-1}$) and complete at ca. 80% saturation (560 $g\,l^{-1}$). Any precipitated material is centrifuged off and additional ammonium sulphate added until no further precipitation occurs.

3.1.2 Precipitation with Trichloroacetic Acid

The proteins are dissolved in water at 15–20 °C and solid trichloroacetic acid is added to a final concentration of ca. 20 $g\,l^{-1}$. Precipitated material is centrifuged off and analysed further.

3.2 Chromatography

3.2.1 Size Chromatography

Due to the low concentrations of cell wall proteins, few fractionation procedures are based on precipitation reactions. Indeed, most of the chromatographic methods are based on high performance (pressure) systems.

As an example of size chromatography, the histidine-rich extensin (HRGP) from *Zea mays*, which eluted at the void volume of a cation exchange resin due to the uncharged nature of the histidine residues at the eluting pH (7.6), could be further fractionated on a Superose-6 gel filtration column. Elution with phosphate buffer showed the presence of at least four components but the HRGP, eluted at 1.8 void volumes, represented 80–90% of the fraction (Kieliszewski et al. 1992).

3.2.2 Ion Exchange Chromatography

Most applications to cell wall (glyco)proteins involve cation exchange chromatography. Two separate applications are cited.

The extensin family, extracted from *Zea mays* (Kieliszewski et al. 1992), was fractionated on a weak cation exchanger, BioRex-70, at pH 7.6 and buffered with 30 mM sodium phosphate. Elution with a 0 to 1 M sodium chloride linear gradient gave the histidine-rich HRGP in the void volume due to the uncharged histidine residues. The threonine-rich HRGP was retarded and eluted at a higher salt concentration. After size exclusion chromatography (see above), further fractionation of the HHRGP was achieved on a strong cation exchange resin, sulphoethyl aspartamide.

The extensins, secreted into the culture medium of suspension-cultured tobacco cells (Kawasaki 1989), have been fractionated by cation exchange chromatography on CM-Toyopearl in the acetate form. Elution was carried out with a linear gradient of 0 to 0.5 M sodium chloride in 50 mM sodium acetate at pH 3.6. While no clear-cut separation of hydroxyproline-containing species was observed, the carbohydrate was clearly present in the more mobile species.

3.3 Electrophoresis

3.3.1 Sodium Dodecylsulphate-Polyacrylamide Gel Electrophoresis (SDS-PAGE)

SDS-PAGE has been used for the fractionation and confirmation of homogeneity of peroxidase isozymes from peanut (*Arachis hypogaea*). The results showed that the cationic forms were localized in the cell walls and the anionic forms in the plasmalemma (Hu et al. 1989). The procedure used was adapted from the method used for many other proteins (Laemmli 1970), but superior results were obtained if prior deglycosylation with trifluoromethane sulphonic acid/anisole was carried out. Since SDS-PAGE primarily separates proteins on the basis of molecular weights, information on size is also achieved. Extensive details of the procedure have been published elsewhere (e.g. Bollag and Edelstein 1991) so only a general guide is given here. The acrylamide content of the gel depends on the molecular size range of the proteins being investigated, 15% being suitable for $15–45 \times 10^3$ Da while only 5% is required for $60–200 \times 10^3$ Da components. The electrophoresis buffer is normally 25 mM Tris - 200 mM glycine containing 0.1% SDS (pH 8.3), but if the gels are to be run at 4 °C, lithium dodecyl sulphate is used. Bromophenol blue is included in the sample compartment as a marker and the gels are run at a constant voltage of 200 V. The resultant amperage depends on the gel thickness being used and will change from 100–110 mA to 60–80 mA during the separation. The gels are subsequently stained with Coomassie Blue for localization of proteins and, if glycoproteins are present, with Schiffs reagent for carbohydrates.

3.3.2 Isoelectric Focusing

This procedure has been used for the fractionation of peroxidase isozymes which had been isolated from different tissue types of tobacco (*Nicotiana tabacum*)

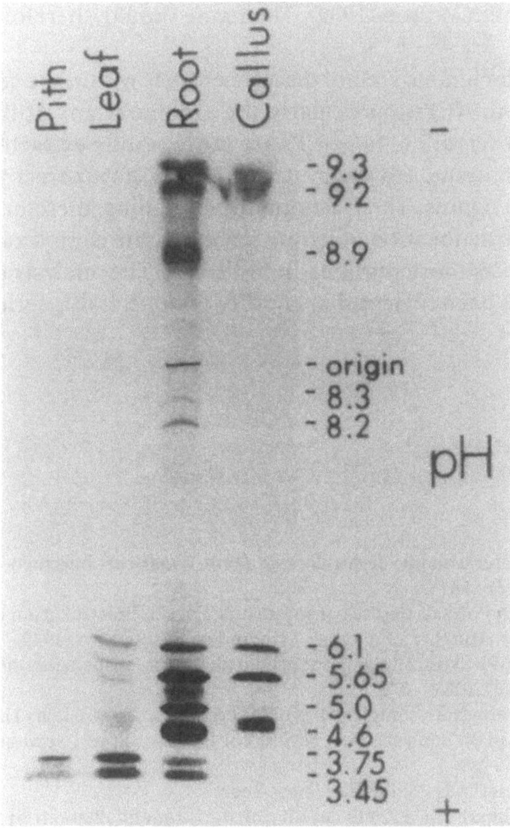

Fig. 4. Isoelectric focusing of peroxidase isozymes from different tobacco tissues. (Lagrimini and Rothstein 1987)

(Lagrimini and Rothstein 1987; Fig. 4). The flat-bed isoelectric focusing was carried out on polyacrylamide gels containing ampholines in the pH range 3.5 to 9.5 for 1.5 h at 0.125 W cm^{-2} and 10 °C. The ampholines are removed by soaking in 10 mM sodium phosphate/150 mM sodium chloride and stained for peroxidase activity with 4-chloro-1-naphthol/H_2O_2.

4 Lignins

Lignins, by weight, represent a major proportion of plant biomass. They are present exclusively in the walls of secondary thickened cells. They are polymers which, using present techniques, cannot be isolated from cell walls without causing extensive alteration/degradation. Hence, there is still debate whether one should refer to lignin or lignins. Since lignins can contain both of the recognized monomer units and different intermolecular bonds are present in most lignins analysed, they are more likely to be a family of related phenylpropenoid copolymers rather than a distinct substance (Monties 1989). The name should, therefore, be lignins.

It is now generally accepted that lignins are not discrete entities within the cell wall. They are covalently bonded to NCPs, particularly the arabinoxylans of the Gramineae. This agrees with the observation that NCPs are more readily extracted after delignification of the specific tissue. However, it is probably not correct to suggest that all NCPs are linked to lignins. The fractionation of lignins, therefore, should really be considered as the fractionation of lignin-carbohydrate complexes. The procedures used are the same as those outlined in Section 2. The analysis of lignin-carbohydrate complexes has been reviewed in another volume of this series (Azuma 1989).

References

Adams GA (1957) Constitution of a water soluble hemicellulose from American beechwood (*Fagus grandifolia*). Can J Chem 35:556–562
Aspinall GO, Hunt K, Morrison IM (1966) Polysaccharides of soy-beans. Part II. Fractionation of hull cell-wall polysaccharides and the structure of a xylan. J Chem Soc Sect C 1945–1949
Aspinall GO, Molloy JA, Craig JWT (1969) Extracellular polysaccharides from suspension cultured sycamore cells. Can J Biochem 47:1063–1070
Azuma J-I (1989) Analysis of lignin-carbohydrate complexes of plant cell walls. In: Linskens HF, Jackson JF (eds) Modern methods of plant analysis. New Series, vol 10. Plant fibres. Springer Berlin Heidelberg New York, pp 100–126
Bollag DM, Edelstein SJ (1991) Protein methods. Wiley-Liss, New York
Brillouet J-M, Joseleau J-P, Utille J-P, Lelièvre D (1982) Isolation, purification and characterization of a complex heteroxylan from industrial wheat bran. J Agric Food Chem 30:488–495
Dean PDG, Johnson WS, Middle FA (1985) Affinity chromatography. IRL Press, Oxford
Erskine AJ, Jones JKN (1956) Fractionation of polysaccharides with copper acetate. Can J Chem 34:820–826
Fry SC (1983) Feruloylated pectins from the primary cell wall: their structure and possible function. Planta 157:111–123
Gaillard BDE (1961) Separation of linear from branched polysaccharides by precipitation as iodine complexes. Nature 191:1296–1297
Gaillard BDE (1965) Comparison of the hemicelluloses from plants belonging to two different plant families. Phytochemistry 4:631–634
Gardner PE, Chang MY (1974) The acetylation of native and modified hemicelluloses. Tappi 57:71–75
Hakomori S (1964) A rapid permethylation of glycolipid and polysaccharide catalyzed by methylsulfinyl carbanion in dimethyl sulphoxide. J Biochem (Tokyo) 55:205–208
Hayashi T, Maclachlan GA (1984) Pea xyloglucan and cellulose I. Macromolecular organization. Plant Physiol 75:594–604
Hayashi T, Marsden MPF, Delmer DP (1987) Pea xyloglucan and cellulose V. Xyloglucan-cellulose interactions in vitro and in vivo. Plant Physiol 83:384–389
Hu CF, Smith R, van Huystee R (1989) Biosynthesis and localization of peanut peroxidases. A comparison of the cationic and anionic isozymes. J Plant Physiol 135:391–397
Jarvis MC, Threlfall DR, Friend J (1977) Separation of macromolecular components of plant cell walls: electrophoretic methods. Phytochemistry 16:849–852
Jones JKN, Stoodley RJ (1965) Fractionation using copper complexes. In: Whistler RL (ed) Methods in carbohydrate chemistry, vol 5. Academic Press, New York, pp 36–38

Josè M, Puigdomènech P (1993) Structure and expression of genes coding for structural proteins of the plant cell wall. New Phytol 125:259–282

Joslyn MA, de Luca GJ (1957) Precipitation of pectin with aluminium hydroxide. Colloid Sci 12:108–130

Kawasaki S (1989) Extensin secreted into the culture medium by tobacco cells I. Purification and some properties. Plant Cell Physiol 30:259–265

Kieliszewski MJ, Kamyab A, Leykam JF, Lamport DTA (1992) A histidine-rich extensin from *Zea mays* is an arabinogalactan protein. Plant Physiol 99:538–547

Keller B (1993) Structural cell wall proteins. Plant Physiol 101:1127–1130

Koleva MI, Achtardjieff C (1973) Isolation of an electrophoretically homogeneous glucomannan using glutaraldehyde-insolubilized concanavalin A. Carbohydr Res 31:142–145

Laemmli UK (1970) Cleavage of structural proteins during the assembly of the head bacteriophage T4. Nature 227:680–695

Lagrimini LM, Rothstein S (1987) Tissue specificity of peroxidase isozymes and their induction by wounding and tobacco mosaic virus infection. Plant Physiol 84:438–442

Maness NO, Mort AJ (1989) Separation and quantitation of galacturonic acid oligomers from 3 to over 25 residues in length by anion-exchange high-performance liquid chromatography. Anal Biochem 178:248–254

McDougall GJ, Fry SC (1991) Purification and analysis of growth-regulating xyloglucan-derived oligosaccharides by high pressure liquid chromatography. Carbohydr Res 219:123–132

Meier H (1965) Fractionation by precipitation with barium hydroxide. In: Whistler RL (ed) Methods in carbohydrate chemistry, vol 5. Academic Press, New York, pp 45–46

Monties B (1989) Lignins In: Dey PM, Harborne JB (eds) Methods in plant biochemistry, Vol 1. Plant phenolics. Academic Press, New York, pp 113–157

Morrison IM (1980) Changes in the lignin and hemicellulose concentrations of ten varieties of temperate grasses with increasing maturity. Grass Forage Sci 35:287–293

O'Dwyer MH (1926) The hemicelluloses. Part IV. The hemicelluloses of beech wood. Biochem J 20:656–664

Scott JE (1960) Aliphatic ammonium salts in the assays of acidic polysaccharides from tissues. Methods Biochem Anal 8:145–197

Scott JE (1965) Fractionation by precipitation with quanternary ammonium salts. In: Whistler RL (ed) Methods in carbohydrate chemistry vol 5. Academic Press, New York pp 38–44

Seth RP, Mukherjee S, Verma SD (1984) Investigation and partial structure of galactomannan A from *Cassia renigera* seed. Carbohydr Res 125:336–339

Stevens BJH, Selvendran RR (1984) Hemicellulose polymers of cabbage leaves. Phytochemistry 23:339–347

Timpa JD (1991) Application of universal calibration in gel permeation chromatography for molecular weight determinations of plant cell wall polymers: cotton fibres. J Agric Food Chem 39:270–275

Valent BS, Albersheim P (1974) The structure of plant cell walls. V. The binding of xyloglucan to cellulose fibres. Plant Physiol 54:105–108

Vlassara H, Brownlee M, Cerami A (1981) Non-enzymatic glycosylation of peripheral nerve proteins in diabetes mellitus. Proc Natl Acad Sci USA 78:5190–5192

Welinder KG, Rasmussen SK, Penel C, Greppin H (1993) Plant peroxidases: biochemistry and physiology. University of Geneva, Geneva

Zitko V, Bishop CT (1965) Fractionation of pectins from sunflowers, sugar beets, apples, and citrus fruits. Can J Chem 43:3206–3214

Isolation and Analysis of Cell Wall Polymers from Olive Pulp

MANUEL A. COIMBRA, IVONNE DELGADILLO, KEITH W. WALDRON, and ROBERT R. SELVENDRAN

1 Introduction

The olive tree (*Olea europaeia*) is native to countries having a Mediterranean climate (Fernández-Díez 1971, 1983). Olive fruit is a drupe, i.e. a meaty stone fruit (Rommani and Jennings 1971), and is usually an elongated spheroid. Depending on the variety, the weight of the fruit may vary from 3 to 10 g. During ripening, the colour changes from green to purple or black. The pulp is very bitter, especially during maturation, and comprises between 70 to 90% of the fruit, the bitter taste is due to the presence of a substantial quantity of phenolic compounds (Amiot et al. 1986; Vlahov 1992). The mature seed is very tough and elongated. The olive fruit is rich in oil which originates from the pulp; most cultivated olives have an oil content of 22% by fresh weight (Kiritsakis and Markakis 1987). The fruit is the raw material for a variety of products including olive oil and table olives. The preparation of table olives is a complex industrial process which involves fermentation and/or alkali treatment (Fernández-Díez et al. 1985). The pulp of the olive fruit is composed of different tissue types (Fig. 1) and there is a paucity of information on the cell walls of these tissues. A knowledge of the structure and chemical composition of olive cell walls is crucial for a better understanding of the biochemical changes that occur in olives during growth, maturation and processing. This chapter describes methods that can be used for the isolation of relatively pure cell walls from olive pulp and the extraction of cell wall polysaccharides by methods that cause minimum degradation. For detailed structural studies, cell wall material virtually free of intracellular compounds is required. However, for certain practical purposes, an alcohol-insoluble residue (AIR) may suffice, provided the limitations of the preparation are borne in mind. The methods used for the separation and characterization of the polymers are described briefly, with particular reference to some of the more recent developments. The experimental details described below are mainly based on our work on the isolation analysis of cell walls of olive pulp (Coimbra 1993; Coimbra et al. 1994).

2 Isolation of Cell Walls from Olive Pulp

In order to isolate relatively pure cell walls from olive pulp, it is necessary to avoid binding or co-precipitation of intracellular compounds such as proteins

Modern Methods of Plant Analysis, Vol. 17
Plant Cell Wall Analysis
Edited by H.-F. Linskens and J.F. Jackson

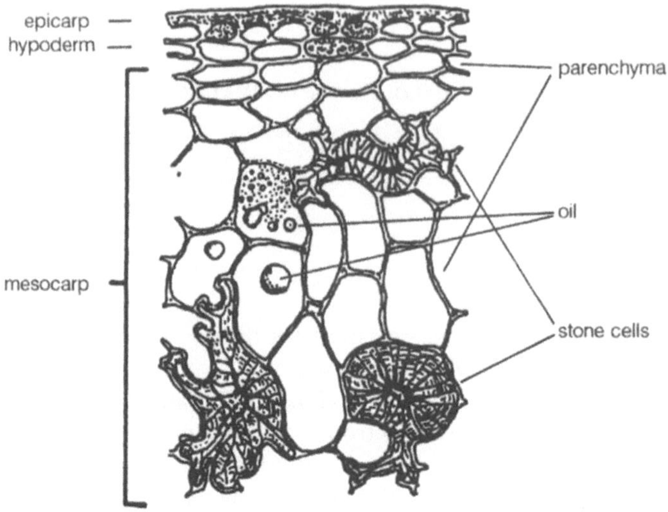

Fig. 1. Olive pulp cell structure. (Winton and Winton 1932)

and phenolics, to remove the olive oil, and to minimize cell wall enzyme activity.

2.1 Preparation and Use of Alcohol-Insoluble Residue (AIR) – General Considerations

Because AIR has been used as a source of cell wall material by various workers, the merits of this preparation will be considered briefly. The preparation of AIR is a relatively easy and quick method for obtaining the cell wall part and has been used in previous studies on olive pulp cell walls (Gil-Serrano et al. 1986; Gil-Serrano and Tejero-Mateo 1988). This method is quite efficient for the inactivation of enzymes while avoiding significant degradation of cell wall polysaccharides. However, the AIR contains much co-precipitated intracellular proteins, nucleic acids, polyphenols and their condensation products with proteins and starch (Newcomb 1963; Loomis and Battaile 1966; Selvendran and O'Neill 1987). Such preparations are of little use for studies on cell wall glycoproteins and proteoglycans (Selvendran 1975a,b; Selvendran et al. 1975 Stevens and Selvendran 1980;), and the co-precipitated compounds can affect the extractabillity of cell wall polysaccharides. The results of methylation analysis of starch-free AIR should be interpreted with caution because of interference from intracellular glycoconjugates, e.g. mannose-rich lectins and cellular enzymes. The dehydration effects of the alcohol may cause interactions between the polymers and the formation of artefacts.

To prepare AIR from olive pulp, olives are stoned, cut into small pieces, immersed in boiling ethanol and homogenized (final concentration 85% v/v aq.)

with an Ultraturrax or equivalent homogenizer (maximum speed, 1 min) and then boiled for 5 min. The mixture is cooled, filtered through a No. 3 glass sinter, and the residue is then extracted twice with chloroform:methanol (2:1, v/v, 3× initial fresh weight) for 30 min under reflux in order to remove the oil. The residue is washed with ether, filtered and air-dried until no variation in weight is detected. The yield of the final residue, completely depleted of oil, is approximately 7% by weight. Unnecessary exposure of cut surfaces of fresh tissue to the air should be avoided; if large quantities of AIR are to be prepared, the pieces of olive pulp may be frozen in liquid nitrogen and then stored below −20 °C until required.

The AIR prepared in this way will be slightly green in colour, and its protein content will be approximately 28%, the majority of which will be of intracellular origin. The AIR will also contain co-precipitated phenolics. The co-precipitated protein and phenolic material are not effectively removed by a solution of phenol:acetic acid:water (PAW 2:1:1, w/v/v), a potent solvent for protein and phenolics (Selvendran 1975a), which only slightly decreases the protein content (28 to 23%). These values for proteins are relatively high when compared with the values proposed for the primary cell walls of dicotyledonous plants (3–6%) (Selvendran 1983). Preliminary studies showed that these co-precipitation effects affected the solubility characteristics of the cell wall polymers. However, the AIR could be used for the determination of component sugars for relevant screening.

Recently, work with cultured tissues of sycamore (Koller et al. 1991) has suggested that the co-precipitated intracellular proteins have a negligible effect on the solubility characteristics of the cell wall polysaccharides. However, this, in our view, is due to the low protein and phenolic content of this residue, and is not generally applicable to material rich in intracellular proteins and phenolics such as olive pulp.

2.2 Preparation of Cell Wall Material (CWM)

In our studies, co-precipitation effects were minimized by using aqueous solvents which have a strong affinity for intracellular compounds, particularly proteins, phenolics and oil. Complete disruption of tissue structure is essential and this is achieved by wet ball-milling the triturated material using the method essencially described by Selvendran and O'Neill (1987). The method, with the relative amounts of material and extractants used, is shown in Fig. 2.

The frozen olive pulp slices are homogenized in a solution of 1.5% sodium dodecylsulphate (SDS) containing 5 mM of sodium metabisulphite ($Na_2S_2O_5$) with an Ultraturrax for 3–5 min. A few drops of *n*-octanol are added to reduce frothing and the $Na_2S_2O_5$ minimizes the formation of oxidation products by phenolics. The triturated material is filtered through a nylon cloth and washed twice with 0.5% SDS solution containing 3 mM of $Na_2S_2O_5$ which removes the bulk of the intracellular material such as phenolics, proteins and nucleic acids. To ensure 'complete' disruption of the tissue structure, the residue is suspended in 0.5% SDS solution and ball-milled (Pascall pot) (60 rpm for 10 h at 1 °C), which gives optimal

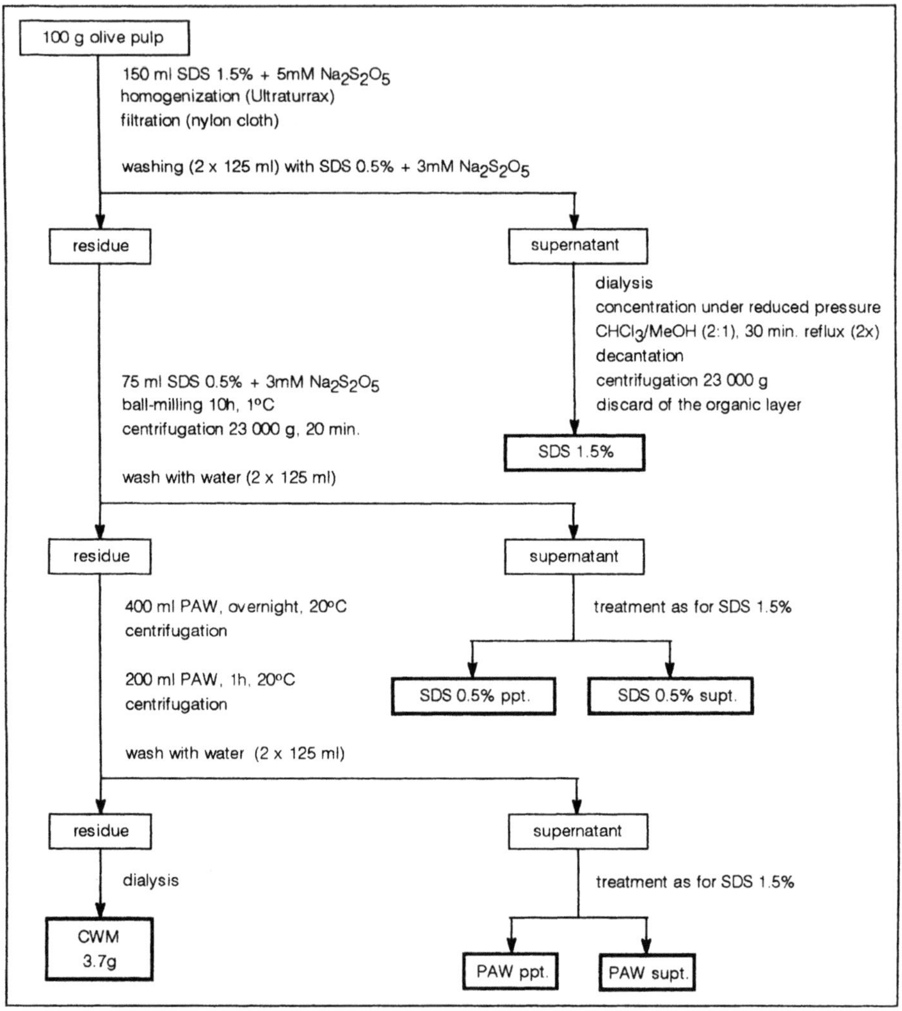

Fig. 2. Preparation of the CWM from olive pulp

cell disruption for olive pulp, and renders the cell contents accessible to the solvents. After ball-milling, the slurry is filtered through a large nylon sieve (3 mm pore size) and the balls washed with distilled water. The material is then centrifuged (23 000 × *g*, 20 min at 1 °C). If the material is ball-milled too finely some difficulty may be experienced in obtaining ready centrifugation. The residue is washed twice with distilled water and the washings are combined with the ball-milled supernatants and retained for the isolation of cold-water-soluble cell wall polymers. The residue is then extracted twice with a solution of phenol:acetic acid:water (PAW 2:1:1, w/v/v); in this treatment, the water retained by the residue should be taken into account and a short treatment with an Ultraturrax blender helps to give a uniform suspension. The residue obtained on centrifugation is washed thoroughly with distilled water (three times) to give a residue virtually free of intracellular protein and phenolic substances. As the olive pulp is free of starch (shown by a negative reaction with I_2/KI), extraction with DMSO is not necessary. In order to remove the residual phenol from the CWM, it is dispersed in water and dialyzed for 3 days at 1 °C with two changes of water daily. A few drops of chloroform and toluene are added to the dialysis flask to minimize microbial growth. The dialyzed residue is then centrifuged to give the purified cell wall material (CWM). The yield of CWM, on a fresh weight basis, is about 3.7% of the olive pulp. The CWM may be kept as a frozen aqueous suspension at –20 °C until required for fractionation studies and an aliquot can be freeze-dried for overall sugar, methylation and spectroscopic analyses.

2.2.1 Material Solubilized During the Preparation of CWM

During the preparation of CWM, small amounts of water- and detergent-soluble cell wall polymers will have been solubilized (Table 1). These polymers can be isolated from the various extracts after dialysis has removed low molecular weight compounds. To ensure complete removal of oil, the concentrated extracts should be treated twice with chloroform:methanol (2:1, v/v) for 30 min, under reflux. From the low carbohydrate content, the 1.5% SDS extracted material can be inferred to be of intracellular origin. However, the material solubilized by 0.5% SDS during ball-milling will contain significant amounts of uronic acid, arabinose, xylose, and galactose. Most of these will have arisen from water-soluble pectic polysaccharides and arabinogalactan proteins. In addition to residual proteins and oil, PAW treatment also solubilizes small amounts of cell wall polymers. Based on

Table 1. Sugar composition of purified cell wall material of fresh olive pulp and material solubilized during purification

Fraction	Yield[a] (g/kg)	Cell wall sugars (mol%)								Total sugar[b] (μg/mg)
		Rha	Fuc	Ara	Xyl	Man	Gal	Glc	Ur.Ac.	
1.5% SDS	6.1 (2.9%)	10	2	30	9	5	15	20	9	119
0.5% SDS										
soluble	0.8 (2.1%)	1	t[c]	25	17	2	10	7	38	663
precipitate[d]	1.1 (1.4%)	2	t	29	9	2	5	25	28	327
PAW										
soluble	0.4 (0.5%)	1	t	19	3	1	3	6	67	338
precipitate[d]	1.0 (2.3%)	2	–	27	15	2	3	30	21	581
Purified CWM	37.1 (90.8%)	1	t	27	12	2	3	29	26	619

[a] Yield is expressed in g/kg fresh weight of olive pulp. The values in parentheses give the carbohydrate content (%) of the fractions.
[b] Values are expressed as μg of "anhydrosugar"/mg dry polymer.
[c] t, Trace.
[d] Materials which precipitated on dialysis.

the carbohydrate composition, the total amount of non-starch polysaccharides solubilized by the above treatments, including 1.5% SDS, has been found to be 6.3% of the dry weight of the purified CWM.

The carbohydrate content of the purified CWM (62%) is relatively small when compared with that of the CWM of onions (Redgwell and Selvendran 1986), potatoes (Ryden and Selvendran 1990a), runner beans (Ryden and Selvendran 1990b) and carrots (Massiot et al. 1988). This, and the observation that the protein content is 9%, suggest that there may be lignin-like material in addition to polysaccharides. The high content of glucose, arabinose and uronic acid in the CWM (Table 1) indicates the presence of cellulose and large amounts of pectic polysaccharides rich in arabinose. The presence of significant amounts of xylose indicates the presence of xylans in addition to xyloglucans.

3 Sequential Extraction of Cell Wall Polymers

3.1 General Considerations

Up to 1985, the procedure most commonly used for the extraction of pectic polysaccharides involved treatment with hot chelating agents such as ammonium oxalate or EDTA (Selvendran et al. 1985). However, these procedures led to significant degradation of the pectic polysaccharides, thereby masking useful information on the structural features of complex pectic polysaccharides. In preliminary studies on the extraction of olive CWM with hot water and hot oxalate at pH 5, we noted the degradation of pectic polysaccharides and the release of small but significant amounts of "free" arabinans. The bulk of similar arabinans and galactans isolated from other plant sources can be inferred to arise from the degradation of more complex pectic polysaccharides rich in arabinose and galactose. The degraded pectic polysaccharides remaining in the "depectinated" residue were subsequently extracted by alkali with the hemicellulosic polysaccharides. Similarly, work on potato CWM (Ryden and Selvendran 1990a) has shown that the bulk of the galactose-rich pectic polysaccharides associated with the cellulosic residue (α-cellulose) is degraded and solubilized by various similar treatments. The sequential extraction described has been designed to cause minimal degradation of the pectic polysaccharides and is based on the procedure described by Selvendran et al. (1985) and Selvendran and O'Neill (1987). The method, the relative amounts of material, and the extractants used are shown in Fig. 3a,b.

3.2 Sequential Extraction

The CWM is sequentially extracted with (1) 50 mM cyclohexane-*trans*-1,2-diamine-NNN'N'-tetra acetate (CDTA), sodium salt, pH 6.5 at 20 °C for 6 h (CDTA-1); (2) 50 mM CDTA, pH 6.5 at 20 °C for 2 h (CDTA-2); (3) 50 mM Na_2CO_3 + 20 mM

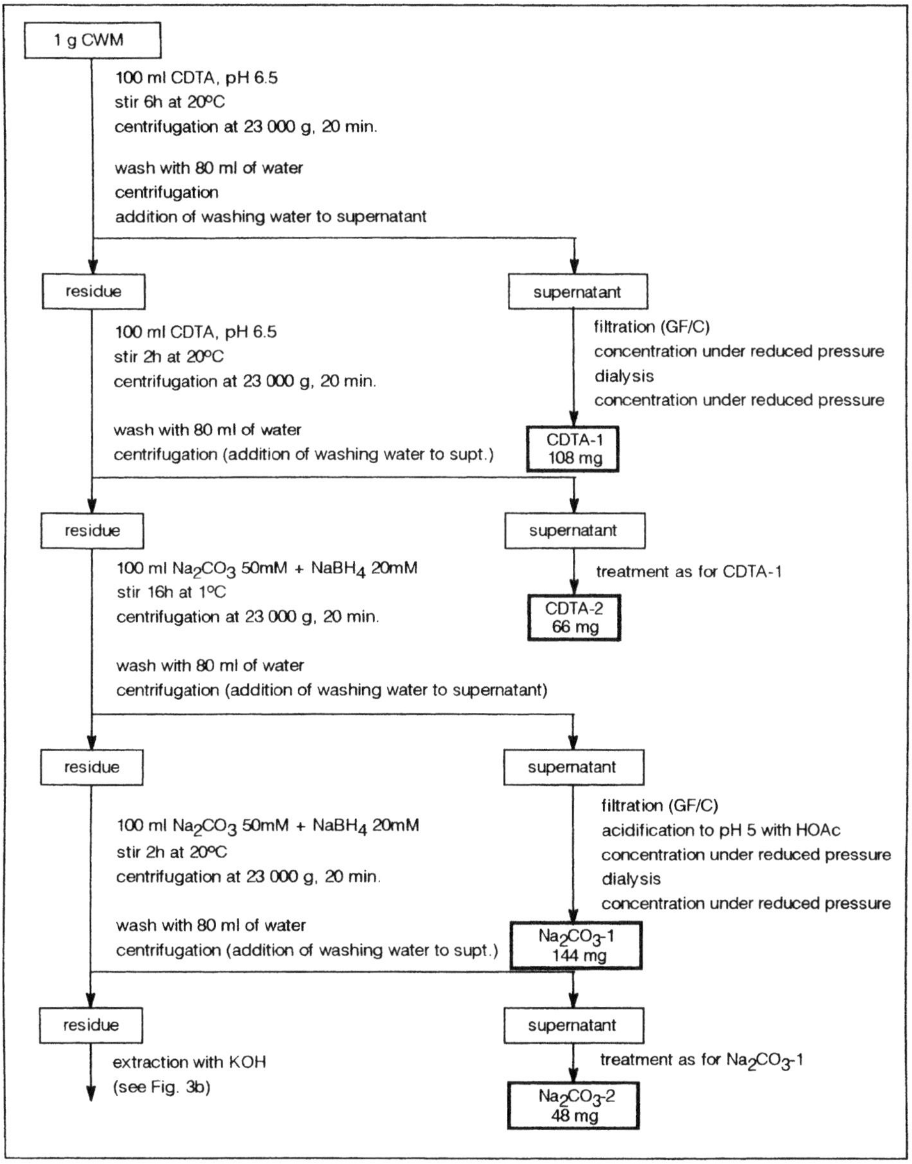

a

Fig. 3a. Sequential extraction of olive pulp CWM. **a** CDTA and Na_2CO_3 extractions. **b** KOH extractions. (Fig. 3b see next page)

$NaBH_4$ at 1 °C for 16 h (Na_2CO_3-1); (4) 50 mM Na_2CO_3 + 20 mM $NaBH_4$ at 20 °C for 3 h (Na_2CO_3-2); (5) 1 M KOH + 20 mM $NaBH_4$ at 1 °C for 2 h; (6) 1 M KOH + 20 mM $NaBH_4$ at 20 °C for 2 h; (7) 4 M KOH + 20 mM $NaBH_4$ at 20 °C for 2 h; and (8) 4 M KOH + 3.5% H_3BO_3 + 20 mM $NaBH_4$ at 20 °C for 2 h. The KOH extractions are carried out with O_2-free solutions under argon. The water used to made these

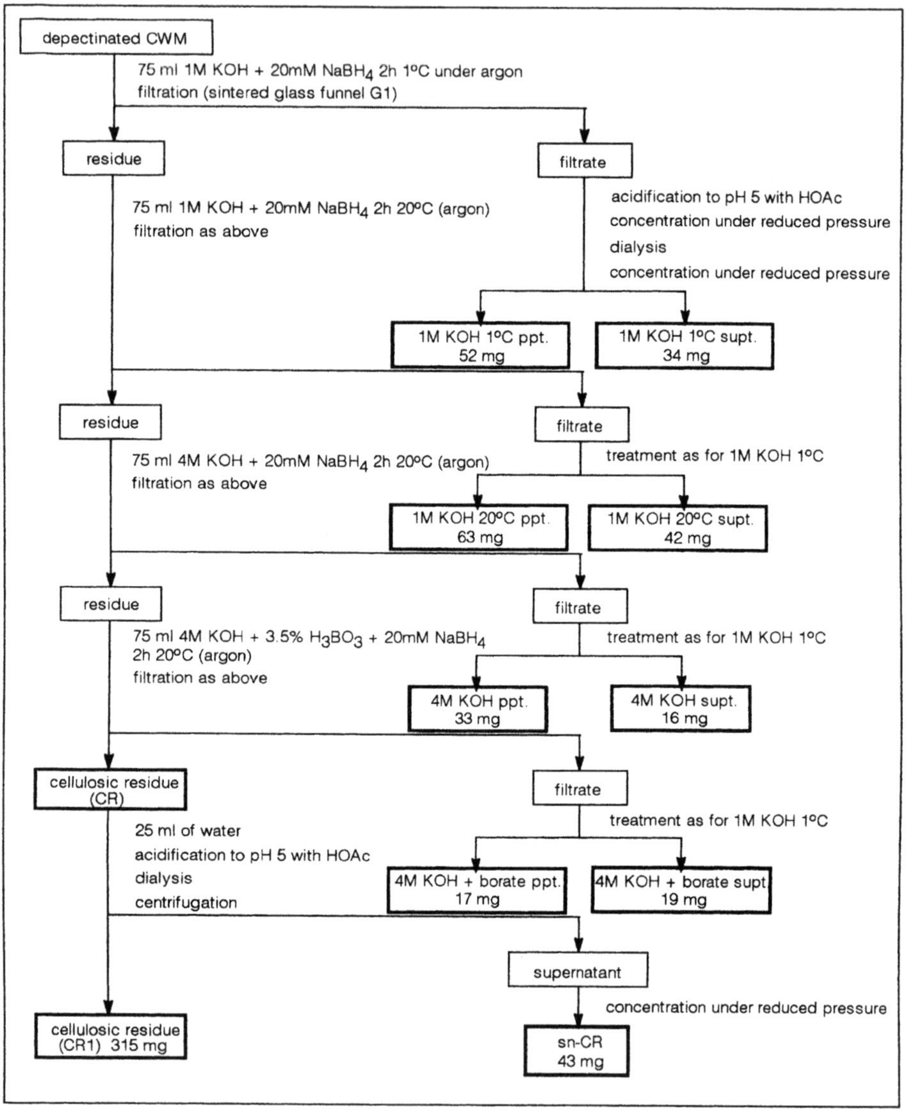

b

Fig. 3b.

solutions is previously boiled for 20 min to liberate all the dissolved oxygen and is allowed to cool under argon. After each extraction, solubilized polymers are separated from the insoluble residue by centrifugation (CDTA and Na_2CO_3 extracts) or by filtration through a G1 glass sinter (KOH extracts). All extracts are filtered through GF/C and dialyzed exhaustively; Na_2CO_3 and KOH extracts are acidified to pH 5 with glacial acetic acid prior to dialysis. Precipitates formed during dialysis of alkali extracts are collected separately. An additional pectic fraction is collected when the cellulosic residue (CR) remaining after the final alkali extraction (4 M KOH + borate) is suspended in water and the solution is acidified to pH 5 and dialyzed. The supernatant of the dialysis of CR (sn-CR) is collected separately by centrifugation from the residue (CR1). All extracts collected after dialysis are concentrated and stored as frozen suspensions at −20 °C or freeze-dried.

From CR1 it is possible to solubilize some additional material by a short warm treatment with chlorite and acetic acid (Jermyn and Isherwood 1956). CR1 is stirred with $NaClO_2$ (0.3%, w/v) and acetic acid (0.12%, v/v) in distilled water at 70 °C for 15 min, according to O'Neill and Selvendran (1980), during which the flask is flushed continuously with argon. The residue is recovered on a G1 glass sinter and the treatment is repeated. The residue is washed thoroughly with distilled water and a final extraction with 1 M KOH + 20 mM $NaBH_4$ at 20 °C for 2 h extracts further material. The chlorite supernatant and water wash are flushed with Ar until the yellow colour fades. This and the alkali extract are filtered and neutralized as for the previous CWM extracts. The final residue (CR2) is suspended in water, neutralized, dialyzed and freeze-dried. The CDTA and alkali extracts after neutralization with acetic acid to pH 5 were extensively dialyzed against several changes of distilled water and freeze-dried for analyses. The yields of polysaccharides and their carbohydrate composition are shown in Table 2.

3.2.1 General Comments to the Method

CDTA is very effective for extraction of pectins from olive pulp CWM. This chelating agent binds calcium ions very effectively at room temperature and solubilizes pectic polysaccharides held in the wall by Ca^{2+}; it is probable that the bulk of these pectic polysaccharides is of middle lamella origin. The carbohydrate recovery of the CDTA soluble polymers was rather low (see Table 2); this was shown to be due to contamination with CDTA, a proportion of which is not readily dialyzable. Mort et al. (1991) also made a similar observation. In our work, the presence of CDTA was shown by ^{13}C-NMR and also by FTIR spectroscopy (Coimbra et al. 1993). From this evidence it is clear that the CDTA in the CDTA extracts cannot be completely removed by dialysis alone. However, experiments in dialyzing the sodium salt of a CDTA solution mixed with Sigma pectin showed that the CDTA could be completely removed by extensive dialysis (K.W.Waldron and R.R.Selvendran, unpubl. results). This would suggest that the carboxyl group of the pectic polysaccharides and CDTA interact with Ca^{2+} to form a "loose" complex from which the CDTA cannot be easily removed. This problem can be overcomed to a significant extent by subjecting the dialyzed CDTA-soluble polymers to graded precipitation with ethanol (described later). In the recent past, Mort et al. (1991) described the use of a solution of imidazole (0.5 M, pH 7) to overcome the problem associated with the presence of CDTA in the pectic polymers. Experiments with orange CWM has shown that the pectic polymers solubilized by the imidazole reagent present more neutral sugars than those solubilized by CDTA (M.J. Barros, M.A. Coimbra and I. Delgadillo, unpubl. results). In our view, the use of CDTA as

Table 2. Sugar composition of fractions of cell wall material of fresh olive pulp obtained by sequential extractions with aqueous solvents

Fraction	Recovery (%)	Cell wall sugars (mol%)								Total sugar[a] (μg/mg)
		Rha	Fuc	Ara	Xyl	Man	Gal	Glc	Ur.Ac.	
CDTA-1	10.8	1	t[b]	11	t	t	1	1	86	753
CDTA-2	6.6	1	t	21	2	t	2	2	71	318
Na_2CO_3-1 (1 °C)	14.4	2	–	40	1	t	3	1	53	767
Na_2CO_3-2 (20 °C)	4.8	1	–	52	1	t	2	1	44	882
1 M KOH 1 °C										
supt.[c]	3.4	1	t	27	31	1	5	13	22	789
ppt.[d]	5.2	1	t	8	67	2	2	7	13	547
1 M KOH, 20 °C										
supt.	4.2	1	t	48	8	9	7	13	14	471
ppt.	6.3	1	t	10	53	2	2	14	18	368
4 M KOH										
supt.	1.6	t	1	15	25	11	10	30	8	936
ppt.	3.3	1	t	41	28	2	2	8	18	134
4 M KOH + borate										
supt.	1.9	1	t	64	5	5	5	8	13	588
ppt.	1.7	1	t	52	12	1	2	9	23	188
CR										
sn-CR	4.3	2	–	53	1	–	3	1	40	936
CR1	31.5	1	–	18	8	1	2	56	14	722
Chlorite/HOAc-sol.	2.6	3	–	56	1	t	4	1	35	732
1 M KOH	4.5	2	t	48	16	2	5	9	18	505
Final residue (CR2)	22.8	1	t	6	7	2	1	78	5	725

[a] Values are expressed as μg of "anhydrosugar"/mg dry polymer.
[b] t, Trace.
[c] Material isolated from the supernatant solution.
[d] Material which precipitated on dialysis.

been well tested on a wide range of plant materials and is, at present, the reagent of choice.

Extraction with sodium carbonate (1 and 20 °C) after extraction with CDTA gives a fairly good yield of pectins (19% of the CWM). This reagent, when used in a very dilute form, at 1 °C and in the presence of sodium borohydride ($NaBH_4$), selectively de-esterifies some pectins. Because the extraction is carried out in the cold, the risk of degradation of the pectic substances by β-elimination is minimized. The Na_2CO_3 extractions solubilize pectic material presumably by hydrolysis of weak ester cross-links. This material is mainly pectic polysaccharides rich in arabinose.

Subsequent extractions with increasing strengths of alkali solubilize polysaccharides presumably by hydrolyzing first the ester linkages, and then by disrupting the hydrogen bonding between the hemicelluloses and the cellulosic microfibrils. Degradation of the hemicelluloses is minimized by using oxygen-free solutions and by reducing latent aldehyde groups with sodium borohydride.

The alkali extracts, on neutralization, give precipitates which can be removed by centrifugation. The precipitates contain predominantly acidic xylans; the polysaccharides in the supernatant are mostly xyloglucans and pectic polysaccharides. The 1 M KOH (1 °C) extraction solubilizes significant amounts of hemicelluloses (e.g. acidic xylans) in addition to pectic polysaccharides, whereas the 4 M KOH extracts are mainly hemicelluloses (e.g. glucomannans and xyloglucans) and only small amounts of acidic polysaccharides. The material solubilized by 4 M KOH + borate (soluble fraction) is very rich in arabinose derived from cell wall glycoproteins. Interestingly, 4 M KOH + borate has been shown to solubilize cell wall glycoproteins from potatoes (Ryden and Selvendran 1990a) and runner beans (Ryden and Selvendran 1990b). The precipitates obtained on dialyzing the material solubilized by KOH solutions have a relatively low carbohydrate content and a significant amount of UV-absorbing material.

The neutralization and dialysis of the residue remaining after 4 M KOH + borate extraction yield a pectic polysaccharide-rich fraction (sn-CR), which accounts for 4.3% of the CWM. It would appear that in the strong alkali the ionization of the $-CH_2OH$ groups on cellulose prevents the diffusion of negatively charged pectic polysaccharides enmeshed within the swollen cellulose matrix. On neutralization of the cellulose-rich suspension, the loss of negative charges on cellulose facilitates the diffusion of the entangled pectic polysaccharides. This phenomenon should be distinguished from the solubilization of xyloglucans bound to cellulose microfibrils by strong alkali, which is predominantly a surface desorption effect.

Studies on cell walls from various plant materials (Ryden and Selvendran 1990a,b) have shown that the cellulose-rich residues (CR) contain cross-linked pectic polysaccharides, wall glycoproteins and phenolic material. Such is the case in olive pulp cell walls. The phenolic material is removed by a short treatment with warm sodium chlorite/acetic acid. It should be noted that this treatment is best applied on the final residue, not earlier, as it causes degradation of acid-labile sugars, certain amino acids in glycoproteins and also of phenolic material (Sjöstrom 1981; Carpita 1984; Ford 1986). Thus, useful information essencially on non-carbohydrate material associated with cell wall polysaccharides may be lost. This is an area which requires closer investigation in future studies to gain a better appreciation of the various types of bonds within the cell wall matrix.

4 Fractionation of the Extracted Cell Wall Polysaccharides

4.1 General Considerations

Various methods including acetone, ethanol and copper acetate precipitations have been used for further fractionation of extracted pectic and hemicellulosic polysaccharides (Jones and Stoodly 1965; Aspinall et al. 1968). Over the last 15–20 years, anion-exchange chromatography has been used for fractionating acidic polysaccharides. Unfortunately, the recovery of polysaccharides from these columns is usually low and is often less than 50%. Some pectic polysaccharides can be quantitatively recovered from DEAE-Trisacryl (Redgwell and Selvendran 1986), but this is not generally applicable to pectic polysaccharides from other sources (e.g. potatoes and runner beans). The factors which result in incomplete recovery of polysaccharides from anion-exchange columns are not fully understood. In our experience, partial fractionation of polysaccharides by graded precipitation with alcohol prior to anion-exchange chromatography have given good recoveries of acidic polysaccharides from wheat bran (Dupont and Selvendran 1987), lignified tissues from runner beans (Selvendran and King 1989) and olive seed hulls (Coimbra et al. 1995a), and, more importantly, the recovery of cell wall polysaccharides of olive pulp (Coimbra et al. 1994).

4.2 Graded Precipitation with Ethanol

The sample is dissolved in water (10 mg/ml), with gentle stirring, overnight if required. The undissolved material is removed by centrifugation and ethanol is added to the supernatant in steps between 10 and 20% to a final concentration of 80% (Selvendran and King 1989). During this period, the mixture is placed at 4 °C for 1 h and the precipitate is removed from the supernatant by centrifugation (23 000 × 2) and the procedure is repeated with the supernatant. In order to remove the ethanol completely, each precipitate is dissolved in water and rota-evaporated at 37 °C. It should be noted that complete removal of the ethanol from the precipitates cannot be achieved by freeze-drying alone. Table 3 (top) shows the effect of the ethanol fractionation of the polymers from the CDTA-1 extract. The FTIR spectra of the 85% ethanol-soluble fraction shows that this fraction contained the majority of the CDTA. Hence, in addition to the partial fractionation of the pectic polysaccharides, this procedure facilitates the removal of the bulk of the CDTA. Analysis of the carbohydrate composition shows that, with the exception of the 85% ethanol-soluble fraction, the polymers of the different fractions have a similar sugar composition. However, they exhibit different degrees of methyl esterification as shown by FTIR spectroscopy. The degree of esterification of the polymers increases with the increase in ethanol concentration. Therefore, it is possible that the degree of esterification is a factor that influences graded precipitation with alcohol. It should be noted that the apparently low recovery of carbohydrate in the 60, 75 and 85% cuts is not due to the presence of contaminating CDTA as shown by FTIR spectroscopy. The poor recovery may be due to incomplete hydrolysis and also precipitation of pectic polysaccharides during the hydrolysis step of analysis.

4.3 Anion-Exchange Chromatography

Anion-exchange chromatography may be used for subfractionation of cell wall polymers, but the recovery of polysaccharides tends to vary depending on the

Table 3. Sugar composition of fractions obtained by ethanol precipitation from CDTA-1 extract (top) and by ethanol precipitation followed by DEAE-Trisacryl chromatography of the Na_2CO_3, 1 °C extract. (bottom)

Fraction	Recovery (%)	Cell wall sugars (mol%) Rha	Fuc	Ara	Xyl	Man	Gal	Glc	Ur.Ac.	Total sugar[a] (µg/mg)
CDTA-1										
EtOH 60% (C1-60)	35.6	2	t[b]	29	t	–	1	1	67	472
EtOH 75%	20.0	3	t	25	t	–	2	1	69	615
EtOH 85%	5.3	2	t	29	1	–	1	2	65	586
EtOH 85% supt.[c]	37.7	32	–	48	7	–	–	–	13	12
Na_2CO_3-1										
Insoluble residue	3.0	2	t	41	1	t	3	3	50	495
EtOH 70%	89.3	2	t	36	t	t	2	1	59	914
Buffer	3.1	2	t	50	8	1	3	15	21	164
B + 0.125 M NaCl	4.9	3	–	71	1	t	4	2	19	581
B + 0.25 M (NC1)	38.2	3	–	68	t	t	3	1	25	804
(NC2)	47.4	2	–	12	1	1	2	1	81	590
B + 0.5 M NaCl	5.2	1	t	17	1	1	2	2	76	605

[a] Values are expressed as µg of "anhydrosugar"/mg dry polymer.
[b] t, Trace.
[c] Supernatant solution.

material being fractionated and also on the type of resin used. For separation of pectic polysaccharides, DEAE-Trisacryl, DEAE-Sephacel, and DEAE-Sephadex have been used by many groups of workers. DEAE-Sephadex tends to shrink during use and this may affect the elution pattern of the polysaccharides. For separation of hemicellulosic polysaccharides such as xyloglucans, DEAE-Trisacryl could be used for the initial fractionation to obtain the bulk of the neutral polysaccharides. Further fractionation can be effected on DEAE-Sephacel using borate buffers. The following comments are based on our experience with cell wall polymers from various edible plants including olives.

A range of anion-exchange columns has been used for the separation of pectic polysaccharides of onions (Redgwell and Selvendran 1986); DEAE-Trisacryl has proved to be the best resin and gives good separation and fractionation of the polysaccharides. This resin has also been used for the purification of pectic arabinans (Joseleau et al. 1983) and has been extensively used for the separation of pectic polysaccharides of olive pulp.

The material is suspended in water (~2 mg/ml) and gently stirred at 20 °C to ensure maximum dissolution; any undissolved material can be removed by centrifugation. The supernatant is diluted with potassium phosphate buffer, pH 6.5, to a final concentration of 50 mM containing 1 mg/ml of polysaccharide material. The solution is passed through a column of DEAE-Trisacryl (1 ml of resin/7.5 µmol uronic acid), phosphate form, at 10 ml/h. The column is then eluted with buffer containing NaCl of increasing strength (0.125, 0.25, 0.5, and 1 M NaCl). Fractions (3 ml) are collected and aliquots (20 µl) are assayed for carbohydrate by the phenol-H_2SO_4 method (Dubois et al. 1956). In this procedure, 20 µl of the fraction is mixed with 80 µl of 5% phenol solution and 0.5 ml of concentrated sulphuric acid, which should be added rapidly onto the liquid surface rather than against the side of the test tube. The mixture is vortexed and boiled in hot water for 5 min and then cooled. The optical density is measured at 490 nm. The appropriate fractions are combined, dialyzed, concentrated, and freeze-dried.

4.3.1 Anion-Exchange Chromatography of Pectic Polysaccharides

DEAE-Trisacryl chromatography has given good separation of the pectic polysaccharides from CDTA and Na_2CO_3-1 extracts, after graded precipitation with ethanol. In the case of the Na_2CO_3-1 extract (70% ethanol cut), the recovery of the polysaccharides from the column was over 95% (Table 3, bottom, and Fig. 4a). In this connection, it should be noted that the fraction between the peaks which

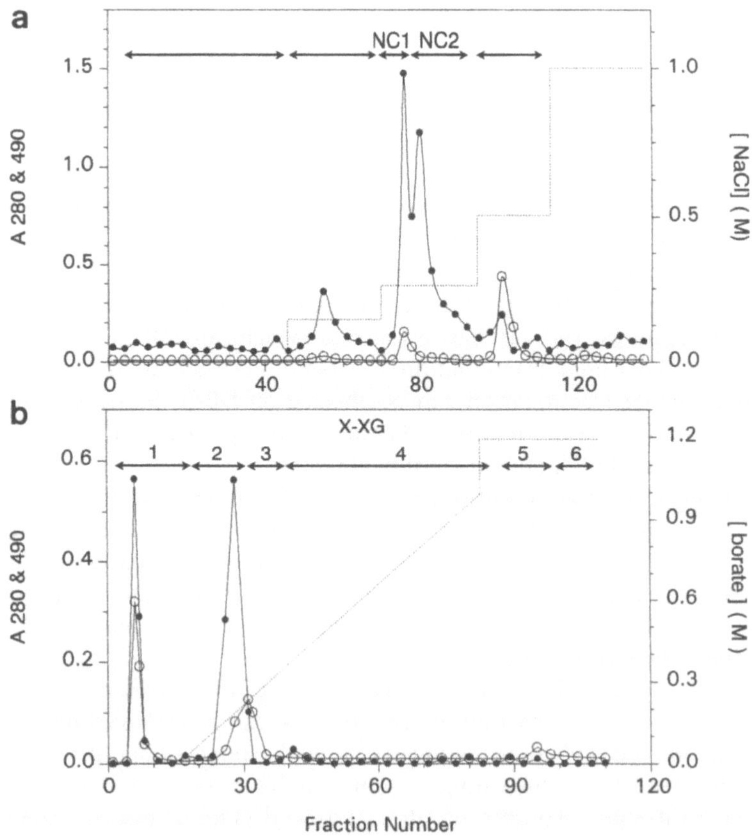

Fig. 4a,b. Anion-exchange chromatography on **a** DEAE-Trisacryl of the pectic polymers from the Na_2CO_3, 1 °C extract (ethanol 70% cut); **b** on DEAE-Sephacel of hemicellulosic polysaccharides from the 1 M KOH, 1 °C extract

exhibit a low carbohydrate profile may contain polysaccharide material and, in some cases, this could amount to up to 15% of the total polysaccharides applied to the column. In this extract, the two main fractions (NC1 and NC2) were shown by sugar analysis to contain different amounts of arabinose and uronic acids (Table 3) and were subjected to further studies such as methylation analysis and ^{13}C-NMR spectroscopy.

4.3.2 Anion-Exchange Chromatography of Hemicellulosic Polysaccharides

The fractions which voided the DEAE-Trisacryl column from the 1 M KOH and 4 M KOH + borate extracts were found to be rich in xyloglucans and hydroxyproline-rich glycoproteins, respectively (Table 4). The xyloglucan-rich fraction was further fractionated with DEAE-Sephacel using borate buffers (Fig. 4b). The use of borate buffers to separate apple xyloglucans was reported by Ruperez et al. (1985). The DEAE-Sephacel is equilibrated with 0.5 M sodium borate buffer, pH 8, over 24 h and is washed overnight with a more dilute solution of buffer (30 mM). The col-

Table 4. Sugar composition of fractions from the 1 M KOH, 1 °C and 4 M KOH + borate extracts

Fraction	Recovery (%)	Cell wall sugars (mol%)								Total sugar[a] (µg/mg)
		Rha	Fuc	Ara	Xyl	Man	Gal	Glc	Ur.Ac.	
1 M KOH, 1 °C supt.[b]										
Insoluble residue[c]	41.5	1	1	54	16	1	4	6	17	158
EtOH 70%	2.0	1	1	17	45	1	6	20	9	861
EtOH 70% supt.	25.0	1	1	15	46	2	6	21	8	796
Buffer (K1A)	62.9	1	1	10	44	2	7	28	7	915
B + 0.125 M (K1B)	17.1	1	–	15	64	t[d]	2	4	14	796
B + 0.25 M NaCl	2.9	2	–	39	20	2	4	10	23	774
1 M KOH, 1°C ppt.										
NaOH insol. res. (K1R)	6.7	1	t	11	11	t	1	74	2	212
NaOH soluble $Cu(OAc)_2$ ppt. (K1Cu)	43.9	1	–	1	91	–	t	1	6	923
pH 10 ppt. (K10)	85.0	1	–	1	87	t	t	2	9	869
pH 10 sol. (K6)	10.0	2	t	3	78	1	1	4	11	821
4 M KOH + bor. supt.										
Buffer (K4A)	21.3	1	t	66	3	10	6	10	4	679
B + 0.125 M NaCl	9.3	2	–	76	2	1	3	4	12	635
B + 0.25 M NaCl	4.7	3	t	62	2	t	4	5	24	561
B + 0.5 M NaCl	1.3	3	1	28	2	38	3	18	7	355

[a] Values are expressed as µg of "anhydrosugar"/mg dry polymer.
[b] Supernatant solution.
[c] Insoluble material obtained on thawing the 1 °C extract.
[d] t, Trace.

umn is packed (1 ml of resin/2.7 μmol uronic acid) and eluted with 1.2 M of buffer during the day and subsequently equilibrated with 30 mM buffer. The material is suspended in water (~5 mg/ml) and gently stirred at 20 °C to ensure maximum dissolution. Any undissolved material can be removed by centrifugation; the supernatant is then diluted with borate buffer, pH 8, to a final concentration of 30 mM containing 1–2 mg/ml of polysaccharide material. The solution is passed through the column of DEAE-Sephacryl, borate form, at 15 ml/h. The column is then eluted with a buffer gradient from 30 mM to 1 M and then 1.2 M. Fractions (3 ml) are collected and aliquots (20 μl) are assayed for carbohydrate by the phenol-H_2SO_4 method (Dubois et al. 1956).

The subfractions from the 1 M KOH extract "neutral" fraction (K1A, Table 4) of olive pulp on the borate ion-exchange column (Fig. 4b) showed a carbohydrate composition, which suggested the occurrence of xyloglucan and xylan complexes. These complexes were further characterized after enzymic digestion with xylanase by fractionation on Sephacryl S-400 (Coimbra et al. 1995b).

Fraction K4A, from 4 M KOH + borate extract (Table 4), was shown to be rich in glycoproteins containing arabinose. These glycoproteins were also found to be rich in hydroxyproline (37 μg/mg).

4.4 Fractionation of Acidic Xylans

From the precipitate obtained on neutralization and dialysis of the 1 M KOH extract, an acidic xylan can be obtained by precipitation with copper acetate (Table 4). The material (100 mg) is dispersed in 1 M NaOH and the insoluble residue is removed by centrifugation; the supernatant (20 ml) is treated with copper acetate solution (7%, w/v; 3.5 ml) and stirred for 3 h at room temperature. The blue precipitate formed is collected by centrifugation and washed with distilled water. This precipitate is dissolved in 5% ethanolic HCl solution (20 ml) and by the addition of water (25 ml) a white precipitate that is formed is collected by centrifugation. The trace of residual acid in the precipitate is neutralized and the suspension dialyzed and freeze-dried to yield about 50 mg of pure acidic xylan. This xylan can be further purified by "pH fractionation" (Coimbra et al. 1994).

5 Chemical and Spectroscopic Analysis

5.1 Neutral Sugars

The bulk of the non-cellulosic polysaccharides can be released by hydrolysis with 1 M H_2SO_4 at 100 °C for 2.5 h. To release the glucose from cellulose quantitatively Saeman hydrolysis conditions (Saeman et al. 1963) have to be used. This involves dispersing the sample (2–3 mg) in 0.2 ml of 72% H_2SO_4 for 2–3 h at 20 °C followed by dilution to 1 M H_2SO_4 (2.2 ml H_2O) and hydrolysis for 2.5 h at 100 °C. It should be noted that both methods of hydrolysis do not quantitatively release the rhamnose associated with pectic polysaccharides. Only about 60–70% of the rhamnose is released; also xylose linked to galacturonic and glucuronic acid (aldobiouronic acid) is not quantitatively released. The neutral sugars released can be isolated and analysed as their alditol acetates (Selvendran et al. 1979). More recently, an improved method of analysis of neutral sugars has been proposed (Blakeney et al. 1983; Harris et al. 1988); this method will be described briefly.

The acid hydrolysate is cooled and an internal standard (2-deoxy-glucose, 1 mg/ml, 200 µl) is added. This solution (1 ml) is then neutralized with 200 µl of solution of 25% NH_3 and the sugars are reduced with 100 µl of 15% $NaBH_4$ in 3 M NH_3 (30 °C, 30 min); the reaction is terminated by two additions of 50 µl of acetic acid. An aliquot (0.3 ml) of this solution is treated with 1-methylimidazole (0.45 ml) and acetic anhydride (3 ml) for 30 min at 30 °C. This solution is then treated with water (3 ml) to decompose the excess acetic anhydride and the acetylated sugars are extracted with dichloromethane (3–5 ml). The dichloromethane phase is washed three times with water and evaporated to dryness; this does not result in any detectable loss of the alditol acetates. The alditol acetates are dissolved in dichloromethane (70 µl) and analysed by GC on an OV-225 capillary column (15 m length, 0.32 mm internal diameter and 0.2 µm film thickness). Capillary columns give good resolution of the peaks and resolve the fucose and rhamnose; the elution time can be as short as 30 min.

5.2 Uronic Acid

Uronic acid is determined colorimetrically by a modification (Selvendran et al. 1979, 1989) of the method of Blumenkrantz and Asboe-Hansen (1973). The sample (2 mg) is hydrolyzed in 0.2 ml of 72% H_2SO_4 for 3 h at room temperature followed by 1 h in 1 M H_2SO_4 at 100 °C. The hydrolysate is filtered through a GF/C filter and diluted ten times with water (pectin samples must be diluted 20 or more times). To 3 ml of ice cold concentrated sulphuric acid containing 50 mM boric acid, 0.5 ml of solution is added, mixed on the vortex and boiled in a water bath for 10 min. After cooling, 100 µl of *m*-phenyl-phenol (0.15% in 0.5% NaOH) is added and the solution is allowed to stand in the dark for 30 min to develop the pink-red colour; the absorbance is read at 520 nm (if the solution is exposed to the light, the external tubes of the rack, more exposed, present less absorbance than the internal tubes, less exposed to the light). Standards can be made with galacturonic or glucuronic acid with a linear correlation from 0 to 40 µg of uronic acid in the tube.

Recently, the uronic acids present in enzymic hydrolysates of cell wall polysaccharides were resolved by chromatography on anion-exchange chromatography on HPLC columns (De Ruiter et al. 1992). This method allows the determination of not only uronic acid but also the carbohydrate composition of the polymers without the need for derivatization.

5.3 Hydroxyproline Estimation

Hydroxyproline (Hyp) was released by 6 M HCl at 110 °C for 24 h and quantified by the method of Kivirikko and Liesmaa as described by Fry (1988). To 400 µl of sample cooled in ice and containing 1–3 µg Hyp is added 400 µl of 0.36% (v:v) bromine solution in 1.25 M NaOH. The mixture is allowed to react for 5 min at 0 °C and 20 µl of 16% Na_2SO_3 is added. The solution is stirred and 400 µl of 5% *p*-dimethylaminobenzaldehyde in 1-propanol is added followed by 200 µl 6 M HCl. The mixture is incubated in a water bath at 95 °C for 2.5 min and in a water bath at 20 °C for 10 min. The absorbance of the solution is determined at 560 nm and the amount of Hyp in the sample is determined by graphic intrapolation using standard solutions of *cis*-4-L-Hyp between 0 and 3 µg/tube.

5.4 Methylation Analysis

The polysaccharides can be methylated by the procedure of Hakomori (1964) or preferably by a modification of the method of Ciucanu and Kerek (1984) and Isogai et al. (1985). Both methods are described and their relative merits are briefly discussed.

In the case of pectic polysaccharides, the polymers to be methylated have to be first de-esterified by a mild alkali procedure as described by Aspinall et al. (1970): the samples are stirred in 0.1 M NaOH + 20 mM $NaBH_4$ for 3–6 h at 1 °C, neutralized, dialyzed, and freeze-dried.

5.4.1 Hakomori Methylation

In this method, the polysaccharides are dispersed in dimethylsulphoxide (DMSO) and activated by treatment with methylsulphinylmethanide anion and methylation with methyl iodide. For the preparation of the anion, see Jansson et al. (1976). The dry polysaccharide (2–3 mg) in dry DMSO (1 ml) is treated with the base sodium methylsulphinylmethanide (1 ml) for 12 h at 20 °C under argon (Jasson et al. 1976). To the ice-cooled mixture, methyl iodide (1 ml) is added dropwise. The mixture is sonicated and left to clear for 3 h at 20 °C. The methylation mixture can be mixed with chloroform:methanol (1 : 1, v/v), filtered and dialyzed against 50% aqueous ethanol (Ring and Selvendran 1978).

5.4.2 Ciucanu and Kerek Methylation

The sample (2–3 mg) is dispersed in 2 ml of dried DMSO and sonicated occasionally until it is fully dispersed. NaOH pellets (100 mg) powdered under argon are added to the solution. The sample is sonicated for 90 min and allowed to stand for a further 90 min. To the solution cool methyl iodide (1 ml) is added and allowed to react 30 min in an ultrasonic bath and left a further 30 min. $CHCl_3$/MeOH (1:1, v/v, 3 ml) is added and the solution is dialyzed against three lots of 50% EtOH. The dialyzate is evaporated to dryness and the methylated material is hydrolyzed with TFA (0.3 ml) at 121 °C for 1 h (Harris et al. 1984), cooled and rotary-evaporated. The partially methylated sugars are then suspended in 0.3 ml of 2 M NH_3 and 20 mg of $NaBD_4$ are added. The mixture is allowed to react at 30 °C for 1 h and the reaction is terminated by the addition for 0.1 ml glacial acetic acid. The acetylation is performed by adding 1-methylimidazole (0.45 ml) and acetic anhydride (3 ml), allowing the mixture to react for 30 min at 30 °C. This solution is treated with water (3 ml) to decompose the excess acetic anhydride and the acetylated sugars are extracted with dichloromethane (3–5 ml). The dichloromethane phase is washed three times with water and evaporated to dryness; this does not result in any detectable loss of the alditol acetates. The partially methylated alditol acetates are dissolved in dichloromethane (70 µl) and analysed by GC on an OV-225 capillay column (30 m length, 0.32 mm internal diameter, and 0.5 µm film thickness with a pre-column of 1 m) and characterized by GC/MS, using the molar response factors of Sweet et al. (1975).

5.4.3 Carboxyl-Reduction of Methylated Polysaccharides

In the case of pectic polysaccharides and glucuronoxylans, the methylated polymers have to be carboxyl-reduced in order to identify the uronic acid derivative. Methylated fractions (2–5 mg) are carboxyl-reduced by a modification of the method described by Lindberg and Lönngren (1978). After methylation and dialysis in 50% ethanol the solvent is evaporated and the completely dried methylated material is allowed to react with $LiAlD_4$ (20 mg) in dried tetrahydrofuran (1 ml) at 65 °C for 4 h. The excess reagent is then destroyed with EtOH (2–3 drops) and water (2–3 drops). The pH is adjusted to neutrality with 2 M H_3PO_3 and 2 ml of $CHCl_3$:MeOH (2:1, v:v) are added. The reduced polymers are removed from the white precipitate by centrifugation and washed thoroughly with $CHCl_3$-MeOH (2:1, v/v). The solvent is evaporated and the hydrolysis, reduction and acetylation proceed as described for neutral sugars. A critique of factors which can lead to undermethylation is discussed in Needs and Selvendran (1993).

5.4.4 Comments on Results of Methylation Analysis

Table 5 shows the results of methylation analysis of the overall CWM, a pectic polysaccharide (NC2), a xylan-xyloglucan complex (K1A), an acidic xylan (K10) and a fraction rich in glycoproteins (K4A). Methylation analysis gives a good indication of the major polymers present. In olive pulp cell walls we can conclude that the major groups of cell wall polymers are:

Table 5. Glycosyl linkage composition of CWM and selected cell wall polysaccharides (values expressed as relative mol%)

Linkage	Fraction					
	CWM	Na_2CO_3 1 °C		1 M KOH 1 °C		4 M KOH + borate
		NC2	NC2[a]	K1A[a]	K10[a]	K4A
T-Rha*p*	–	2.3	1.6	–	–	–
2-Rha*p*	1.6	1.6	1.7	–	–	0.4
3-Rha*p*	0.1	0.7	0.8	0.2	0.2	–
2,3-Rha*p*	–	0.3	–	–	–	–
2,4-Rha*p*	0.5	1.2	1.6	–	–	0.3
3,4-Rha*p*	–	1.3	1.0	–	0.2	–
T-Fuc*p*	0.1	0.5	0.5	0.3	–	–
T-Ara*f*	13.2	11.9	11.0	6.1	0.3	16.1
2-Ara*f*	0.6	–	0.3	0.4	–	20.3
3-Ara*f*	5.0	6.0	5.2	0.7	–	12.2
5-Ara*f*	12.7	17.1	13.3	2.6	–	7.3
2,3-Ara*f*	–	2.4	–	–	–	–
2,5-Ara*f*	–	0.5	0.4	–	–	–
3,5-Ara*f*	12.6	18.0	14.2	2.6	–	6.9
Arabinitol	0.9	2.2	0.6	0.2	–	0.5
T-Xyl*p*	1.5	1.0	0.8	3.4	0.4	0.4
2-Xyl*p*	3.3	–	–	7.6	–	0.7
4-Xyl*p*	6.0	3.5	2.4	23.3	85.1	1.3
2,3-Xyl*p*	0.8	–	–	–	–	–
2,4-Xyl*p*	0.4	–	–	8.5	12.5	–
Xylitol	2.1	–	0.6	0.8	0.5	–
T-Man*p*	–	–	–	–	–	0.3
4-Man*p*	2.3	–	–	2.1	–	10.4
4,6-Man*p*	0.7	–	–	0.5	–	3.1
T-Gal*p*	–	1.5	1.2	4.0	–	3.7
2-Gal*p*	0.3	–	–	–	–	0.7
3-Gal*p*	–	0.4	0.3	–	–	0.3
4-Gal*p*	0.3	3.7	3.5	–	–	1.4
6-Gal*p*	0.1	–	–	–	–	–
3,4-Gal*p*	0.2	–	0.4	–	–	–
3,6-Gal*p*	–	–	–	–	–	0.7
4,6-Gal*p*	0.2	1.3	–	–	–	–
T-Gal*p*A	–	–	1.6	–	–	–
4-Gal*p*A	–	–	33.3	–	–	–
3,4-Gal*p*A	–	–	0.7	–	–	–
T-Glc*p*	0.3	–	–	0.3	0.4	0.2
4-Glc*p*	27.2	6.1	3.0	9.0	–	10.0
2,4-Glc*p*	0.2	1.3	–	0.5	–	–
3,4-Glc*p*	0.1	–	–	0.5	–	–
4,6-Glc*p*	2.8	0.9	–	21.2	–	2.8
Glucitol	3.9	14.4	–	1.7	0.2	–
T-Glc*p*A	–	–	–	3.5	3.8	–

[a] Sample carboxyl-reduced with $LiAlD_4$.

- extractable pectic polysaccharides;
- pectic polysaccharides associated with the cellulosic residue;
- xylan-xyloglucan complexes;
- acidic xylans;
- hydroxyproline-rich cell wall glycoproteins;
- cellulose.

5.5 ^{13}C-NMR Studies on Cell Wall Polysaccharides

The following comments are restricted to the occurrence of special structural features shown by ^{13}C-NMR. For example, unambiguous evidence was obtained for the occurrence of terminally linked β-arabinofuranose residues (T-β-Ara*f*) in the pectic polysaccharides examined. Figure 5 and Table 6 show the results of ^{13}C-NMR spectral studies with two of the pectic polysaccharides isolated from a Na_2CO_3-soluble extract. Unlike the pectic polysaccharides of tomato, in which the galacturonic acid of the pectin examined did not show significant signals, the Na_2CO_3-soluble pectic polysaccharides of olives showed significant signals for the galacturonic acids of the pectins. This is similar to the observation of sugar-beet pectin (Keenan et al. 1985), which shows the signals of the galacturonic acid molecules.

Spectra (100.4 MHz) were recorded under conditions of broad-band proton decoupling in a deuterated water solution at 40 °C (Ryden et al. 1989). Chemical shifts can be compared with other ^{13}C-spectral data: Joseleau et al (1977, 1983), Capek et al. (1983) and Swamy and Salimath (1991) for "arabinans" and Pressey and Himmelsbach (1984), Keenan et al. (1985) and (Ryden et al. 1989) for pectins.

5.6 Fourier Transform Infrared Spectroscopy (FTIR)

This technique seems to be very good to determine the presence of CDTA in CDTA extracted samples and to evaluate the degree of esterification of the polysaccharides. Recently, it has been shown that microscopy FTIR can be used to

Table 6. Peak assignments of ^{13}C-NMR spectra of fractions NC1 (Fig. 5a) and NC2 (Fig. 5b)

Residue	Chemical shift (δ)					
	C-1	C-2	C-3	C-4	C-5	C-6
T-α-L-Ara*f*	109.8	83.5	78.8	86.1	63.4	–
T-β-L-Ara*f*	103.7	78.6	76.6	84.3	65.3	–
(1→3)-α-L-Ara*f*	109.4	83.1	84.8?	84.6?	63.4	–
(1→5)-α-L-Ara*f*	109.8	83.5	79.0	84.6	69.2	–
(1→3,5)-α-L-Ara*f*	109.4	83.1	84.6	83.8	68.8	–
(1→4)-α-D-Gal*p*A	101.4	70.4	71.0	80.3	73.3	178

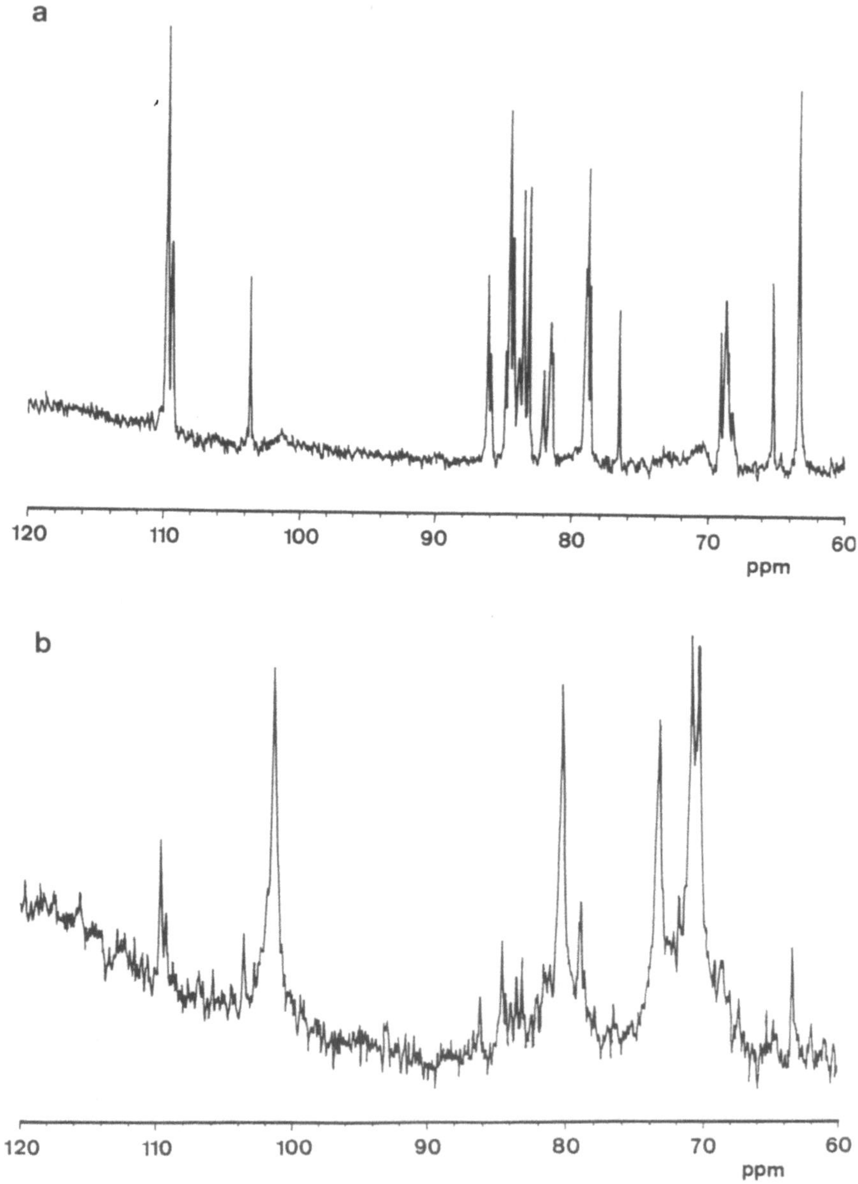

Fig. 5a,b. ^{13}C-NMR spectra of pectic polysaccharides from Na_2CO_3, 1 °C extract: **a** NC1; **b** NC2

follow the sequential extraction of the polysaccharides from CWM within the "fingerprint" range 1200–900 cm^{-1} (McCann et al. 1992). The relative amount of neutral sugars present in pectic polysaccharides as well as the ratio xylose/glucuronic acid in glucuronoxylans can be quickly assessed using FTIR (M.A. Coimbra and I. Delgadillo, unpubl. results). FTIR spectra were obtained at a resolution of 8 cm^{-1}. Polymers were incorporated into KBr (spectroscopic grade) and pressed into a 1-mm pellet. Spectra were recorded at the absorbance mode from 2000 to 800 cm^{-1}.

6 Concluding Remarks

In this chapter, we have described some of the recent developments in cell wall analyses of soft tissues with particular references to its applicability in the isolation analysis of cell walls from olive pulp. The olive pulp, as indicated, is a very difficult material to work with because of its high oil, protein and phenolics content. The preparation of relatively pure cell walls is crucial for any detailed fractionation studies, however, trends in development, maturation and ripening of the fruit can be obtained using the AIR. With such residues care must be taken in interpreting the results of methylation analysis, particularly from mature fruit as the xylose (and cellulose) residues of the lignified sclereids might be incompletely methylated due to the presence of lignin.

We were able to isolate relatively pure polysaccharides from the purified CWM. These polysaccharides could be further examined using various analytical methods. The presence of pectic polysaccharides in the 0.5% SDS extract deserves comment because these polysaccharides are loosely held in the wall and can be isolated in appreciable amounts with minimum contamination with intracellular compounds. The significance of these polysaccharides is not clear.

We hope that the improved methods described in this chapter will pave the way for new and exciting studies on olives, fresh and processed.

References

Amiot M-J, Fleuriet A, Macheix J-J (1986) Importance and evolution of phenolic compounds in olive during growth and maturation. J Agric Food Chem 34:823–826

Aspinall GO, Craig JWT, Whyte JL (1968) Lemon-peel pectin. I. Fractionation and partial hydrolysis of water-soluble pectin. Carbohydr Res 7:442–452

Aspinall GO, Cottrell IW, Molloy JA, Uddin M (1970) Lemon-peel pectin. III. Fractionation of pectic acids from lemon-peel and lucerne. Can J Chem 48:1290–1295

Blakeney AB, Harris PJ, Henry RJ, Stone BA (1983) A simple and rapid preparation of alditol acetates for monosaccharide analysis. Carbohydr Res 113:291–299

Blumenkrantz N, Asboe-Hansen G (1973) New method for quantitative determination of uronic acids. Anal Biochem 54:484–489

Capek P, Toman R, Kadosova A, Rosik J (1983) Polysaccharides from the roots of the marsh mallow (*Althaea officinalis* L.): structure of an arabinan. Carbohydr Res 117:133–140

Carpita NC (1984) Fractionation of hemicelluloses from maize cell walls with increasing concentrations of alkali. Phytochemistry 23:1089–1093

Ciucanu I, Kerek F (1984) A simple and rapid method for the permethylation of carbohydrates. Carbohydr Res 131:209–217

Coimbra MA (1993) Isolation and characterisation of cell wall polysaccharides from olive. PhD Thesis, University of Aveiro, Portugal (in Portuguese)
Coimbra MA, Delgadillo I, Waldron KW, Selvendran RR (1993) Cell wall pectic polysaccharides of olive pulp. Anais da Faculdade de Ciências do Porto, University of Oporto, Portugal (in press)
Coimbra MA, Waldron KW, Selvendran RR (1994) Isolation and characterisation of cell wall polymers from olive pulp (*Olea europaea* L.). Carbohydr Res 252:245–262
Coimbra MA, Waldron KW, Selvendran RR (1995a) Isolation and characterisation of cell wall polymers from the heavily lignified tissues of olive seed hull. Carbohydr Polymers (in press)
Coimbra MA, Rigby NM, Selvendran RR, Waldron KW (1995b) Investigation of the occurrence of xylan-xyloglucan complexes in the cell walls of olive pulp (*Olea europaea*). Carbohydr Polymers (in press)
De Ruiter GA, Schols HA, Voragen AGJ, Rombouts FM (1992) Carbohydrate analysis of water-soluble uronic acid-containing polysaccharides with high-performance anion-exchange chromatography using methanolysis combined with TFA hydrolysis is superior to four other methods. Anal Biochem 207:176–185
Dubois M, Gilles KA, Hamilton JK, Rebers PA, Smith F (1956) Colorimetric method for determination of sugars and related substances. Anal Chem 28:350–356
DuPont MS, Selvendran RR (1987) Hemicellulosic polymers from the cell walls of beeswing wheat bran. Part I Polymers solubilised by alkali at 2 °C. Carbohydr Res 163:99–113
Fernández-Díez MJ (1971) The olive. In: Hulme AC (ed) The biochemistry of fruits and their products, vol 2. Academic Press, London pp 255–279
Fernández-Díez MJ (1983) Olives. In: Reed G (ed) Food and feed production with microorganisms. In: Rehm H-J, Reed G (eds) Biotechnology – a comprensive treatise in 8 volumes, vol 5. Verlag Chemie, Weinheim, pp 379–397.
Fernández-Diez MJ Ramos RC, Fernández AG, Cancho FG, Pellisó FG, Vega MN, Herédia-Moreno A, Mosquera IM, Navarro LR, Quintana MCD, Roldan FS, Garcia PG, Gomez-Millan AC (1985) Biotecnologia de la aceituna de mesa. Consejo Superior de Investigaciones Cientificas, Instituto de la Grasa y sus Derivados, Madrid, Seville
Ford CW (1986) Comparative structural studies of lignin-carbohydrate complexes from *Digitaria decumbens* (Pangola grass) before and after chlorite delignification. Carbohydr Res 147:101–117
Fry SC (1988) The growing plant cell wall: chemical and metabolic analysis. Longman, Harlow, Essex, UK
Gil-Serrano A, Tejero-Mateo P (1988) A xyloglucan from olive pulp. Carbohydr Res 181:278–281
Gil-Serrano A, Mateos-Matos MI, Tejero-Mateo MP (1986) Acidic xylan from olive pulp. Phytochemistry 25:2653–2654
Hakomori SI (1964) A rapid permethylation of glycolipid, and polysaccharide catalyzed by methylsulphinyl carbanion in dimethyl sulfoxide. J Biochem 55:205–208
Harris PJ, Henry RJ, Blakeney AB, Stone BA (1984) An improved procedure for the methylation analysis of oligosaccharides and polysaccharides. Carbohydr Res 127:59–73
Harris PJ, Blakeney AB, Henry RJ, Stone BA (1988) Gas chromatographic determination of the monosaccharide composition of plant cell wall preparations. J Assoc Off Anal Chem 71:272–275
Isogai A, Ishizu A, Nakano J (1985) A new facile methylation method for cell-wall polysaccharides. Carbohydr Res 138:99–108
Jansson PE, Kenne L, Liedgren H, Lindberg B, Lönngreen J (1976) Chemical communications. University of Stockholm, Sweden, pp 1–76
Jermyn MA, Isherwood FA (1956) Changes in the cell wall of the pear during ripening. Biochem J 64:123–132
Jones JKN, Stoodley RJ (1965) Fractionation using copper complexes. In: Whistler RL (ed) Methods in carbohydrate chemistry, vol 5. Academic Press, New York, pp 36–38
Joseleau J-P, Chambat G, Vignon M, Barnoud F (1977) Chemical and ^{13}C NMR studies on two arabinans from the inner bark of the young stems of *Rosa glauca*. Carbohydr Res 58:165–175
Joseleau J-P, Chambat G, Lanvers M (1983) Arabinans from the horsebean (*Vicia faba*). Carbohydr Res 122:107–113

Keenan MHJ, Belton PS, Matthew JA, Howson SJ (1985) A ^{13}C NMR study of sugar-beet pectin. Carbohydr Res 138:168–170
Kiritsakis A, Markakis P (1987) Olive oil: a review. Adv Food Res 31:453–482
Koller A, O'Neill MA, Darvill AG, Albersheim PA (1991) A comparison of the polysaccharides extracted from dried and non-dried walls of suspension-cultured sycamore cells. Phytochemistry 30:3903–3908
Lindberg B, Lönngren J (1978) Methylation analysis of complex carbohydrates: general procedure and application for sequence analysis. Methods Enzymol 50:3–33
Loomis WD, Battaile J (1966) Plant phenolic compounds and the isolation of plant enzymes. Phytochemistry 5:423–438
Massiot P, Rouau X, Thibault J-F (1988) Characterisation of the extractable pectins and hemicelluloses of the cell wall of carrot. Carbohydr Res 172:229–242
McCann MC, Hammouri M, Wilson R, Belton P, Roberts K (1992) Fourier transform infrared microspectroscopy is a new way to look at plant cell walls. Plant Physiol 100:1940–1947
Mort AJ, Moerschbacher BM, Pierce ML, Maness NO (1991) Problems encountered during the extraction, purification, and chromatography of pectin fragments, and some solutions to them. Carbohydr Res 215:219–227
Needs PW, Selvendran RR (1993) An improved methylation procedure for the analysis of complex polysaccharides including resistant starch and a critique of the factors which lead to undermethylation. Phytochemical Analysis 4:210–216
Newcomb EH (1963) Cytoplasm-cell wall relationships. Annu Rev Plant Physiol 14:43–64
O'Neill MA, Selvendran RR (1980) Glycoproteins from the cell wall of *Phaseolus coccineus*. Biochem J 187:53–63
Pressey R, Himmelsbach DS (1984) ^{13}C NMR spectrum of a D-galactose-rich polysaccharide from tomato fruit. Carbohydr Res 127:356–359
Redgwell RJ, Selvendran RR (1986) Structural features of cell-wall polysaccharides of onion *Allium cepa*. Carbohydr Res 157:183–199
Ring SG, Selvendran RR (1978) Purification and methylation analysis of cell wall material from *Solanum tuberosum*. Phytochemistry 17:745–752
Ruperez P, Selvendran RR, Stevens BJH (1985) Investigation of the heterogeneity of xyloglucans from the cell walls of apple. Carbohydr Res 142:107–113
Ryden P, Selvendran RR (1990a) Structural features of cell-wall polysaccharides of potato (*Solanum tuberosum*). Carbohydr Res 195:257–272
Ryden P, Selvendran RR (1990b) Cell-wall polysaccharides and glycoproteins of parenchymatous tissues of runner bean (*Phaseolus coccineus*). Biochem J 269:393–402
Ryden P, Colquhoun IJ, Selvendran RR (1989) Investigation of structural features of the pectic polysaccharides of onion by ^{13}C NMR spectroscopy. Carbohydr Res 185:233–237
Saeman JF, Moore WE, Millet MA (1963) Sugar units present. In: Whistler RL (ed) Methods in carbohydrate chemistry, vol 3. Academic Press, New York, pp 54–69
Selvendran RR (1975a) Analysis of cell wall material from plant tissues: extraction and purification. Phytochemistry 14:1011–1017
Selvendran RR (1975b) Cell wall glycoproteins and polysaccharides of parenchyma of *Phaseolus coccineus*. Phytochemistry 14:2175–2180
Selvendran RR (1983) The chemistry of plant cell walls. In: Birch GG, Parker KJ (eds) Dietary fibre. Applied Science, London, pp 95–147
Selvendran RR, King SE (1989) Structural features of the cell wall polysaccharides of the parchment layers of the pods of mature runner beans. Carbohydr Res 195:87–99
Selvendran RR, O'Neill MA (1987) Isolation and analysis of cell walls from plant material. In: Glick D (ed) Methods of biochemical analysis, vol 32. Wiley (Tube and Science), New York, pp 25–153
Selvendran RR, Davies AMC, Tidder E (1975) Cell wall glycoproteins and polysaccharides of mature runner beans. Phytochemistry 14:2169–2174
Selvendran RR, March JF, Ring SG (1979) Determination of aldoses and uronic acid content of vegetable fiber. Anal Biochem 96:282–292
Selvendran RR, Stevens BJH, O'Neill MA (1985) Developments in the isolation and analysis of cell walls from edible plants. In: Brett CT, Hillman JR (eds) Biochemistry of plant cell walls. SEB seminar 28, Cambridge Univ Press, London, pp 39–78

Selvendran RR, Verne AVFV, Faulks RM (1989) Methods of analysis of dietary fibre In: Linskens HF, Jackson JF (eds) Modern methods of plant analysis, New Series vol 10. Springer, Berlin Heidelberg, New York 234–259

Sjöstrom E (1981) Wood chemistry: fundamentals and applications. Academic Press, London

Stevens BJH, Selvendran RR (1980) The isolation and analysis of cell wall material from the alcohol-insoluble residue of cabbage (*Brassica oleracea* var. capitata). J Sci Food Agric 31:1257–1267

Swamy NR, Salimath PV (1991) Arabinans from *Cajanus cajan* cotyledon. Phytochemistry 30:263–265

Sweet DP, Shapiro RH, Albersheim P (1975) Quantitative analysis by various GLC response factor theories for partially methylated and partially ethylated alditol acetates. Carbohydr Res 40:217–225

Vlahov G (1992) Flavonoids in three olive (*Olea europaea*) fruit varieties during maturation. J Agric Food Chem 58:157–159

Winton AL, Winton KB (1932) Structure and composition of foods, vol 1. Wiley, New York

Determination of Cell Wall Autolysis

PEDRO P. GALLEGO

1 Introduction

During plant development, an important metabolism occurs which includes cell wall turnover (synthesis de novo and degradation) of cell wall polysaccharides mediated by a wall-associated multi-enzyme system. The term autolysis pertains to the processes of the cell wall, in which both enzymes and polysaccharides intervene as catalysts.

This mechanism causes a loosening of the cell wall that is an integral step for cell wall elongation and, consequently, plant growth. In addition, cell wall autolysis occurs extensively during fruit ripening - being responsible for softening - and also during abscission and senescence processes.

Two different approaches have been used to estimate indirectly the activities of wall-bound enzymes: (1) in vitro cell wall autolysis consisting of the isolation of the active cell wall and following by incubation in an appropriate buffer, and (2) assays using wall-associated enzymes extracted and purified from active cell walls and incubated with inactive cell walls (or cell wall fractions). Extraction and partial purification of cell wall-binding enzymes have been included in some reports of in vitro autolysis experiments (Huber and Nevins 1981b; Llamazares et al. 1987; Labrador and Nevins 1989; Gallego and Zarra 1994), but since this procedure is a collateral aspect and not really a feature of autolysis, no further details are given here.

Cell wall autolysis in vitro is a useful and standardized technique applied to the study of changes in cell wall polysaccharides mediated by multi-enzyme system activity. Also, it provides a good system where the complex enzyme-substrate remains fairly intact when compared with methods involving cell-protein extractions.

The first report of autolysis in vitro (Lee et al. 1967) showed that isolated cell walls derived from corn were capable of solubilizing polysaccharides in the incubation medium. Since then, a large number of reports describing cell wall autolysis in algae, fungi and higher plants have been documented (Table 1).

In vitro cell wall autolysis has been used to indirectly estimate the activity of enzymes involved in in vivo cell wall loosening (Dopico et al. 1991; Muñoz et al. 1993b). Consequently, young tissues show a greater autolysis capacity than older tissues with a lower growth rate capacity (Labrador and Nicolas 1985; Seara et al. 1988; Hoson 1990). Also, autolysis assays were widely employed during fruit growth (Gallego 1992) and ripening (Rushing and Huber 1987; Huber and Lee

Modern Methods of Plant Analysis, Vol. 17
Plant Cell Wall Analysis
Edited by H.-F. Linskens and J.F. Jackson

Table 1. Examples of plant material used in cell wall autolysis studies

Plant material	References
Algae	
Chlamydomonas spp.	Schlösser (1981)
Chlorella fusca	Walter and Aach (1987)
	Loos and Meindl (1985)
Chlorella ellipsoidea	Satoh and Takeda (1989)
	Araki and Takeda (1992)
Fungi	
Mucor hiemalis	Ohno et al. (1986)
Sporobolomyces ruberrinus	Shiraishi and Fujii (1987)
Clostridium sterilium	Bronnenmeier and Standenbauer (1988)
Gymnosperms	
Pinus pinaster	Llamazares et al. (1987)
Angiosperms	
Monocots	
Zea mays	Huber and Nevins (1979)
Oryza sativa	Labrador and Nicolas (1982)
Dicots	
Pisum sativum	Labrador and Nicolas (1985)
	Dopico et al. (1986)
Glycine max	Nari et al. (1986)
Cicer arietinum	Revilla et al. (1985)
	Seara et al. (1988)
Gossypium hirsutum	Bucheli et al. (1987)
Vigna angularis	Hoson (1990)
Lycopersicon esculentum	Rushing and Huber (1984)
	Gallego et al. (1991)
Actinidia deliciosa	Gallego and Zarra (1995)

1988; Gallego et al. 1991) to characterize the metabolic processes that occur during fruit softening, as well as the related enzymes.

The autolysis procedure involves several steps, including: (1) isolation of active cell walls; (2) isolation of inactive walls; (3) incubation in an appropriate buffer; and (4) analysis of autolysis products. However, before describing in above steps in detail, it is very important to analyse some factors involved during cell wall isolation and incubation because these can seriously alter and, consequently, falsify the autolysis results. These factors are principally buffer, pH and ionic strength, depending on the plant material being investigated.

2 Factors Altering Cell Wall Isolation and Autolysis

Exposure of cell walls to chelating buffers, pH extremes and strong ionic environments must be avoided. Since the cell wall is a polyanion, these factors alter wall-charge distribution, promoting modifications in the structure of cell wall polysaccharides and in wall-bound enzyme environments. Likewise, pretreatments of cell walls by heat, phenolic compounds or some organic solvents (i.e. methanol: chloroform) may modify and inhibit the enzyme activities associated with cell walls (see Sect. 4 for details).

2.1 Plant Material

Cell wall autolysis has been described widely throughout the plant kingdom. As mentioned above, the plant materials used in autolysis experiments in vitro are summarized in Table 1.

Cell wall composition and structure, as well as wall-associated enzymes, depend on the stage of development studied and the plant material. With regard to the stage of development and the different parts of the plants, it has been demonstrated that cell walls prepared from different tissues showed different capacities for autolysis of polysaccharides than others (Labrador and Nicolas 1985; Seara et al. 1988; Hoson 1990). There was also a dramatic increase in vitro cell wall autolysis activity as fruit ripening proceeded (Rushing and Huber 1984). With regard to the plant material in monocots, the main sugar released by autolysis was glucose (Lee et al. 1967; Huber and Nevins 1979; Labrador and Nicolas 1982; Bucheli et al. 1987); in dicots, the main sugars were galactose and arabinose (Labrador and Nicolas 1985; Hoson 1990; Muñoz et al. 1993a); in fruits, uronic acids (Rushing and Huber 1984; Gallego and Zarra 1995) were described.

This means that different enzymes are involved in these autolysis experiments and thus different approaches must be carried out in each case, in order to optimize the autolysis conditions in each case (see Sect. 5.1).

In general, plant tissues must be used immediately after collection, or frozen at −20 °C to avoid the occurrence of wall autolysis prior to cell wall isolation procedures.

2.2 Buffers

Cold distilled water has been used as aqueous medium for homogenization in both growing plants (Labrador and Nicolas 1982) and fruits (Rushing and Huber 1987). However, buffers present an obvious advantage: it is possible to control the pH of the homogenization medium.

The most widely used buffers in autolysis experiments contain citrate and/or phosphate: e.g. citrate phosphate (Huber and Nevins 1981a; Dopico et al. 1986; Llamazares et al. 1987; Muñoz et al. 1993a) and Na-phosphate (Huber and Nevins 1979; Revilla et al. 1985; Hoson 1990). However, buffers containing both citrate (Carr and Ng 1959; Keijbets et al. 1976; Buescher and Hobson 1982) and phosphate (Fry 1988) must be avoided, especially in dicots and fruit tissues with a high pectin content, due to their chelation properties which enhance both enzymic and non-enzymic pectin solubilization.

With this in mind and the previously noted disadvantage of these buffers, the use of 20–40 mM Na-acetate (Rushing and Huber 1984; Huber and Lee 1988; Koch and Nevins 1989), more commonly 50 mM (Rushing and Huber 1987; Nock and Smith 1987; Jackman et al. 1992; Gallego et al. 1992), has been proposed.

In conclusion, buffers with chelating properties or containing chelating agents or salts with high ionic strength must be avoided.

2.3 pH

A great number of results suggest that both enzymic activities and wall-enzyme binding are pH-dependent. Studies to determine the cell wall pH in vivo indicated that it may be close to 5 (Schopfer 1989). Likewise, the majority of wall hydrolases also have an optimun pH of around 5 (Taiz 1984), with the exception of pectin methylesterase, which has an optimun pH at 8 (Goldberg 1984). Therefore, the use of buffers (Geigy Scientific Tables 1984) with a pH range between 4.5 and 5.5 is recommended during cell wall extractions and in vitro autolysis experiments. It should be borne in mind, however, that the pH can be altered during and after the homogenization process, depending on the material used (i.e. fruits can modify the pH of the homogenization buffer due to their high organic acid content). Therefore, during homogenization and incubation procedures, the pH should be checked.

The use of an extreme pH must be completely avoided, because strongly acidic (Selvendran 1975; Hohl et al. 1991) or alkaline (Kiss 1974; Gross 1984) treatments may modify (β-elimination and/or de-esterification of pectins) or release certain polymers from the cell walls. It has also been shown that alkaline pHs (over 6.5) favour polygalacturonase solubilization (Pressey 1986).

2.4 Ionic Strength

Addition of some salt solutions increase the ionic strength of the buffer and enhance the results obtained by in vitro cell wall autolysis. The maintenance of a certain ionic strength is necessary to avoid adsorption by cell walls and also to provide an adequate environment for cell wall-bound enzymes.

Concentrations of NaCl in the range of 150–300 mM have been reported as optimal when the activity of purified polygalacturonase isoenzymes (Pressey and Avants 1973) or tomato fruit autolysis experiments were carried out (Rushing and Huber 1987). Moreover, increasing the ionic strength of the cell wall suspension tends to shift the pH profile towards lower pH values (Nari et al. 1986) and enhances autolysis release (Rushing and Huber 1987).

Some reports, however, have established that the use of salt solutions with a high ionic strength can solubilize wall-associated enzymes (Hobson 1964; Ali and Brady 1982) and inhibit cell wall autolysis (Labrador and Nicolas 1985). Thus, concentrations of 600 mM NaCl or higher inhibited purified polygalacturonase activities (Pressey and Avants 1982) as well as cell wall autolysis (Rushing and Huber 1987). When autolytically active walls were pretreated with 1–3 M NaCl (Huber and Nevins 1979; Rushing and Huber 1987) or 1–12 M LiCl (Huber and Nevins 1979; Huber and Nevins 1981b), a great reduction in the capacity to liberate sugars was also found.

In conclusion, high pH (over 7) as well as high ionic strength (salt concentrations over 1 M) must be avoided because of the enhanced extraction of some hydrolases from cell walls (above all in fruits). The use of buffers in a range of 10–50 mM, containing 150 mM NaCl (at least in fruit experiments) is recommended.

3 Isolation of Active Cell Walls

Several methods are described for the isolation and purification of cell wall materials. All of them have advantages and disadvantages, thus it is best to use a combination. Efforts must be made to minimize degradation of polymers and cell wall enzyme activity during the isolation procedures.

Isolation of cell walls in organic solvents or aqueous buffers has been widely proposed. However, it has been reported (Selvendran et al. 1985; Fry 1988) that during cell wall isolation in organic solvents (mainly 70–80% ethanol) intracellular proteins, among other substances, (starch, low-M_r sugars, etc.), may bind to walls, both by co-precipitation and ionic binding. These intracellular proteins would then seriously interfere with the subsequent autolysis experiments.

Both co-precipitation and ionic binding effects can be avoided by using aqueous buffers during cell wall isolation. However, their use could have several disadvantages if carried out improperly, such as loss of water-soluble components and wall autolysis. This can be avoided if cell wall isolation is preformed in aqueous buffers at or below 4 °C.

In general, plant material is homogenized in cold aqueous medium, filtered (rather than centrifuged), washed with 1 l of cold buffer, followed by washing with cold organic solvents. It should be also be noted that all solutions and glassware should be chilled to 4 °C. The best results are obtained when all homogenization procedures are carried out in the cold room. Many variations in this basic scheme have been reported.

The recommended method with the relative amounts of plant materials, buffers, pHs, etc. and solvents is shown in Table 2.

3.1 Homogenization in Aqueous Medium

Cell walls are obtained by homogenization of the tissues in cold aqueous medium, at a ratio of 1 : 5 (g/ml) for young growing plants and 1 : 2 (g/ml) for fruits. The amount of tissue varies depending on plant material: 5–30 g for seedlings (Huber

Table 2. Procedure for the isolation of active cell walls

Plant material: Use immediately after collection or freeze at –20 °C until used.
Step 1. Usually 30 g of growing plant material (i.e. coleoptiles, epicotyls, etc.) is placed in cold Na-acetate (50 mM, pH 5) and homogenized in a Polytron set at maximum speed for 3 min at a ratio of 1:5 (g/ml) or 100 g of fruit pericarp is homogenized in the same buffer containing 150 mM NaCl at a ratio of 1:2 (g/ml).
Step 2. Homogenate is filtered through a Miracloth and washed with 1 l of cold buffer, followed by washing with cold 80% ethanol and acetone.
Step 3. Active cell walls are air-dried at 25 °C and stored at –20 °C in a tightly stoppered bottle containing silica gel.

and Nevins 1979; Revilla et al. 1985; Llamazares et al. 1987) and 100–200 g for fruits (Rushing and Huber 1984; Gallego and Zarra 1995).

A large variety of combinations of buffers and pH values have been utilized. However, as mentioned previously, the selection of an appropriate buffer is critical for the autolysis experiments. Once an adequate buffer has been selected, the type of plant material should be taken into account since seedlings and fruits have different optimum pHs and ionic strengths within the range proposed.

The process of homogenization is usually carried out on an Omminixer (Huber and Nevins 1989) and Waring blender (Rushing and Huber 1984; Koch and Nevins 1990), however, the use of a Polytron has been widely proposed (Gross and Wallner 1979; Fry 1988; Huber 1991; Gallego and Zarra 1995). Homogenization on a cool pestle and mortar (Bucheli et al. 1987) should be carried out with caution since samples must be very finely ground so that they are as homogeneous as possible. In all cases, some drops of octanol can be applied to avoid the formation of foam during the homogenization procedure.

3.2 Filtration and Washes

The homogenate should be transferred to a coarsely fritted glass filter (Huber and Nevins 1979; Bucheli et al. 1987) or to a Miracloth (Rushing and Huber 1984; Labrador and Nicolas 1985; Nock and Smith 1987) and washed with 1 l of cold buffer. Subsequently, one of the following two procedures can be applied:

1. Wash with 1 l of cold ethanol (80–95%) in order to retain only the polysaccharide fraction, wash then with 1 l acetone at –20 °C to eliminate any residue of aqueous medium, and air dry at 25 °C (Huber and Lee 1988; Gallego et al. 1991). Air-dried cell walls are usually stored at –20 °C in a tube with cap and placed in a tightly stoppered bottle containing silica gel.
2. Wash directly with 1 l acetone at –20 °C and 1 l of cold buffer (Huber and Nevins 1979; Labrador and Nicolas 1985; Dopico et al. 1986; Seara et al. 1988). Note that this procedure requires that isolated cell walls must be used immediately because storage of cell walls in aqueous medium could activate the enzymes and

alter the properties of the cell wall components. If not used immediately, the last cold buffer wash should be avoided.

Both methods are accurate, giving similar results if done correctly.

4 Isolation of Inactive Cell Wall (Controls)

Cell walls incubated in aqueous medium are able to release part of their own material via a non-enzymic process. Therefore, it is important to have a good control which allows the differentiation between enzymic and non-enzymic sugar releases from cell wall autolysis.

In contrast to active cell walls, samples to be used as controls inactive cell walls require the use of methods that inactivate the enzymes and which involve a different or additional treatment. These treatments cause modifications in the cell wall structure of controls. Consequently, the use of inactivating methods which minimize, as far as possible, modifications in the cell wall structure is important. In any case, the comparison of sugars released by control and active cell wall samples should be made with caution.

Most procedures recommend the use of high temperature to inactivate cell wall-associated enzymes. As an alternative to heat inactivation treatments, solutions containing phenol (a powerful protein extractant and denaturant) have been proposed. Finally, autolysis incubation at low temperature avoids the use of differential treatments for control and sample.

4.1 Boiled in Hot Water

Cell walls are commonly inactivated by boiling in water for 10–15 min; they are used as controls in in vitro autolysis of seedlings (Labrador and Nicolas 1982; (Revilla et al. 1985; Dopico et al. 1986; Nari et al. 1986) and fruits (Pressey and Avant 1982; Rushing and Huber 1984). Unfortunately, this is not an appropriate method for pectin-rich materials since hot water (90–100 °C) promotes the solubilization of pectins via β-elimination (Selvendran et al. 1985).

4.2 Boiled in Alcohol

Heat inactivation in the presence of ethanol may protect the pectin polymers from degradation via β-elimination. Knee (1973) proposed boiling in 80% ethanol to inactivate enzymes without pectin degradation. In in vitro autolysis experiments, both methanol (Llamazares et al. 1987; Hoson 1990) and more extensively 80% ethanol (Rushing and Huber 1984, 1987; Huber and Lee 1988; Jackman et al. 1992) were employed as the inactivation method.

However, there is a strong controversy regarding the use of the alcohol-heat inactivation method. Some authors (Huber and Lee 1988) showed that refluxing in 80% ethanol did not completely inactivate polygalacturonase bound to alcohol-insoluble isolates from tomatoes, or cell walls used in in vitro autolysis. It has also been suggested that treatment of potato cell walls with methanol:chloroform (Sasaki et al. 1987) or boiling 80% ethanol promoted irreversible conformational changes or aggregation of pectin polysaccharides (Sasaki and Nagahashi 1989).

Moreover, Selvendran (1975) showed that treatment of cell walls with 80% ethanol promoted the co-precipitation not only of proteins, but also of low molecular weight components (pectin substances).

In contrast, Koch and Nevins (1989) demonstrated that homogenization of the cell wall in ethanol (50–95%), followed by heating at 80 °C for 20 min, completely inactivated polygalacturonase, however, in this case, cell walls were also treated with methanol:chloroform.

With all this in mind, if the cell walls are boiled in ethanol for use as controls, special precautions should be taken when using plant materials with a high content of pectin polysaccharides and polygalacturonase (i.e. fruits) until further in vitro autolysis studies are available that clarify the situation.

4.3 Phenol-Acetic Acid-Water

The mixture phenol-acetic acid-water (PAW, 2:1:1, w/v/v) has been included in protocols for the isolation of cell walls because it effectively removes proteins from extracted cell walls (Selvendran 1975; Jarvis 1982; Fry 1988). This method was successfully employed to remove polygalacturonase and pectin methylesterase completely from tomato pericarp (Seymour et al. 1987a). Another additional advantage is its inability to release structural polysaccharides (Selvendran et al. 1985).

However, when PAW was used in studies of enzymically active cell walls (Rushing and Huber 1984; Huber and Lee 1988), this treatment did not completely eliminate pectin solubilization under autolytic conditions. Also, Huber (1991) showed that PAW removes 50% or more of cell wall calcium, resulting in increased pectin solubility. Moreover, Huber (1992) demonstrated that the use of PAW does not achieve complete inactivation of wall-associated pectin hydrolase activity. As an alternative, inactivation of wall-bound enzymes by exposing cell walls to phenol buffered at neutral pH was recommended (Huber 1991).

4.4 Buffer Phenol at pH 7

Cell walls inactivated by exposure to buffer phenol at pH 7 have been recently used in cell wall autolysis studies (Gallego et al. 1991; Huber 1992; Gallego and Zarra 1995).

However, although buffered phenol is recommended instead of PAW and heat treatments (Huber 1991), more studies must be carried out to determine whether other cell wall parameters not yet examined could be negatively affected by this phenolic solution. In fact, depolymerization of hemicelluloses from corn cell walls inactivated by phenol buffer after 24h of incubation has recently been reported (Peña et al. 1993).

In conclusion, phenolic solvents are good protein extractants, but poor structural polysaccharide extractants. It should also be pointed out that they may promote changes in some cell wall features, not discussed in detail here. Thus, if used, both control cell walls and active cell walls must be analysed.

4.5 Low Temperatures

Cell wall autolysis experiments using temperatures of 4 °C (Rushing and Huber 1984; Koch and Nevins 1989) and 0 °C (Gallego and Huber, unpubl.) have been described.

An obvious advantage of low temperature is that it does not involve any differential pretreatment of controls to inactivate cell wall enzymes which may alter their cell wall structure. Our results (Gallego and Huber, unpubl.) suggest that wall-associated enzymes are inactivated at 0 °C during in vitro autolysis experiments, the amount of sugar released being similar to that from inactivated cell walls (boiled in 80% ethanol and treated with buffered phenol) at 0 °C. Also, Koch and Nevins (1989) suggested that the amount of sugars solubilized by cell wall autolysis at 4 °C shows the extent of hydrolysis occurring during the aqueous phase of the wall isolation procedure. However, more studies are being carried out to demonstrate the existence or non-existence of enzymic activities during in vitro autolysis at 0 °C and to characterize the nature of products released and the cell wall residue.

5 Autolysis Incubations

Several buffers with a wide range in molarity and pH have been described in autolysis experiments. However, as discussed in Section 2, the process of autolysis incubation is dependent upon plant material, buffers, pH, ionic strength of the medium and other variables, mainly temperature and period of incubation.

Most procedures recommend incubations at 34–37 °C over a period of 24–30 h. However, lower temperatures, e.g. 25 °C (Llamazares et al. 1987; Revilla et al. 1985) and 30 °C (Bucheli et al. 1987), as well as shorter incubation periods, e.g. 1–1.5 h (Jackman et al. 1992), 6–8 h (Lee et al. 1967; Rushing and Huber 1987), or as long as 48 h (Huber and Nevins 1981a) and 7 days (Araki and Takeda 1992), have also been described.

Experiments to determine the optimum pH for autolytic experiments reveal that the maximum autolytic capacity was achieved in the range of pH 5–6: for algae, pH 5 (Araki and Takeda 1992), gymnosperms, pH 5.5–6.5 (Llamazares et al. 1987), monocots, pH 5.5 (Labrador and Nicolas 1982) and dicots, pH 5–5.5 (Revilla et al. 1985; Rushing and Huber 1984). Other results (Nari et al. 1986) showed that the rate of autolysis at pH 5–5.5 is enhanced after incubating the cell walls at pH. 8. These results are supported by the idea that pectin methylesterase (a wall-bound enzyme with a pH optimum of 8; Goldberg 1984) promotes the generation of negative charges in the cell wall, thus decreasing the local pH which facilitates autolysis (Moustacas et al. 1986). These results agree with in vitro degradation of isolated tomato cell walls by polygalacturonase which is enhanced by the addition of pectin methylesterase (Pressey and Avants 1982; Seymour et al. 1987b; Koch and Nevins 1989).

Also, cell wall autolysis is enhanced by the addition of 150–300 mM NaCl to the incubation buffer (Rushing and Huber 1987), but this has only been largely adopted in fruit autolysis studies (Huber and Lee 1988; Gallego and Zarra 1995).

5.1 Optimization of the Autolysis Reaction

As previously discussed, it is necessary to establish the optimum conditions for the in vitro autolysis experiments for each material to be examined. Autolysis assays can be characterized by the kinetic curves of cell walls released in an appropriate buffer, where some other factors (period of incubation, temperature, pH, etc.) are varied.

The optimization of the autolysis procedure, in order to determine the optimum values of the aforementioned factors, can be realized depending on the variable that we intend to characterize first. The general experimental procedure is explained in section 5.2. One example, which we recommend, is as follows:

1. Incubate the cell walls at different pHs using 50 mM Na-acetate buffer (range 3.8–5.6) or Tris buffer (range 5.2–8.6) for 24 h at 34 °C. Most of the autolysis products are released during this period of time.
2. With the pH value already determined, select the optimum temperature (a shorter incubation time will be sufficient, e.g. 6 h).
3. Finally, characterize the kinetics of the autolytic products released at different incubation times with the previously determined optimum pH and temperature.

This basic scheme can be varied with regard to the order in which the optimum values of the different parameters are determined.

5.2 General Procedure

Cell walls are water-insoluble, thus care must be taken to achieve complete hydration. Some drops of ethanol added prior to the buffer help to disperse cell wall powder. Moreover, performance is improved when suspensions are incubated with shaking.

Generally, active and inactive cell walls are suspended in 50 ml glass centrifuge tubes hydrated thoroughly in the reaction buffer and incubated in a shaking or magnetically stirred water bath.

During long periods of incubation, antimicrobial substances such as 0.02% sodium azide and 0.01% Thimerosal are commonly added to avoid contamination. However, these products can interfere with some colorimetric methods (see Sect. 6.1). Alternatively, some drops of toluene or 0.05% chlorbutol can be used.

During optimization and characterization processes, aliquots from the suspension can be collected at intervals during autolysis incubation; they are then filtered through a glass-fiber filter in vacuum and analyzed immediately or frozen at −20 °C to avoid soluble-enzyme activities. After aliquot collection, the same amount of fresh buffer must be added to the incubation medium to maintain a constant weight/volume ratio. If there is a residue, it should be combined with the fresh buffer and returned to the incubation medium.

Once the incubation conditions (pH, temperature, time, etc.) are established, suspensions are filtered through a glass-fiber filter and the residue is washed with fresh buffer and resuspended or washed in cold acetone. The residue-acetone mixture is filtered again through a previously weighed glass-fiber filter and dried to a constant weight for cell wall dry weight determinations. Both the cell wall residue and the solution results of combining all supernatants are analyzed as described in Section 6. The recommended procedure and conditions of incubation are summarized in Table 3.

Table 3. Procedure for autolysis incubation

Material: Active and inactive cell walls (controls)[a].
Step 1. Cell walls (20–50 mg) are suspended in 50 ml glass centrifuge tubes and hydrated in cold Na-acetate (50 mM, pH 5[b]) at a ratio of 4:1 (mg/ml). In fruit cell walls buffer should contain 150 mM NaCl.
Step 2. Suspensions are incubated in a shaking water bath at 34 °C[b] over a period of 24 h[b].
Step 3. During optimization, aliquots (ca. 250 µl) should be collected at intervals, filtered through a glass-fiber filter (Whatman GF/A). Fresh buffer (250 µl) must be returned to the incubation medium to maintain the w/v ratio. If there is a residue, it must be combined with the fresh buffer and returned to the medium. Aliquots should be immediately analysed or stored at −20 °C.
Step 4. Final suspensions are filtered through glass-fiber filter (Whatman GF/A) and washed with 1 vol of fresh buffer. Resulting supernatant should be either analysed or frozen.
Step 5. Residue is washed in 50 ml cold 100% acetone and air-dried. Dry cell wall weight should be determined.
Step 6. Cell wall residue and supernatant are analysed as described in Section 6.

[a] Active cell walls can be used as controls by incubating under the same conditions, but at 0 °C, or active cell walls can be inactivated as described in Section 4.
[b] These values are purely symbolic. Proceed as described in Section 5.1 to characterize the optimum pH values, temperature and incubation time.

6 Analysis of Autolysis Products

Cell wall autolysis in vitro promotes both a quantitative release of cell wall material in the incubation medium, and qualitative effects, i.e. structural changes in wall polysaccharides. Most autolysis reports focus on the description of the products released during autolysis experiments, but the study of residues (Nock and Smith 1987) has not been widely reported. Efforts to elucidate the importance of these qualitative effects on cell walls by autolysis analysis must be intensified. Furthermore, the analysis of both solubilized and residual products from the inactive cell walls is of particular importance, providing information not only about the effects of incubation on the cell wall, but also the effectiveness of the inactivation method employed.

6.1 General Methods

Here, only the analysis of the solubilized product is reviewed; the characterization of residues is described in other chapters.

The total sugar content is usually determined by the phenol-sulphuric method (Dubois et al. 1956). Other methods using anthrone for hexoses (Scott and Melvin 1953) and orcinol for pentoses (Dische 1962) can also be employed.

Uronic acid content can be detected by carbazole (Galambos 1967) and more commonly by m-hydroxybiphenyl (Blumnenkrantz and Asboe-Hansen 1973) methods. The first method, however, is subject to interference by neutral sugars, while the second presents more sensitivity and minimal interference.

Moreover, any antimicrobial substances contained in the samples may interfere with all the above methods (especially sodium azide with anthrone and carbazole). In some cases, it can be resolved if a standard curve is made containing these solutions.

For reducing sugar measurements, the Nelson-Somogy method (Somogy 1952) is most often proposed. Several other methods, including ferrocyanide 2,4,6-tripypdyl-s-triacine (Avigad 1975); 2-cyanoacetamide (Honda et al. 1982) and hexaronic acids (Milner and Avigad 1967), have been also used in autolysis experiments (Rushing and Huber 1984; Nock and Smith 1987; Gillet et al. 1992).

Neutral sugars can be analysed by the hydrolysis of sugar samples with 2 N trifluoroacetic acid for 1 h at 121 °C and converted to alditol acetates as described by Albersheim et al. (1967). The amount of acetylated sugars is determined using gas liquid chromatography.

6.2 Ethanol Precipitation

Commonly, products from either aliquots or final samples of autolysis incubations are filtered through a glass-fiber filter and precipitated by adding ethanol to a final concentration of 80%. The solution is maintained at −20 °C for at least 1 h and filtered through a glass filter or centrifuged at 10 000 *g* for 30 min. It is then washed with 1 vol of 80% ethanol and filtered or centrifuged in the same manner as above. The supernatant constitutes the oligomeric fraction and the pellet the polymeric fraction.

6.3 Gel Permeation Chromatography

To study the size and nature of the products released during autolysis different types of columns, usually Bio Gel P, Sephadex or Sepharose, have been employed.

Bio Gel P-2 is the column most often used because it permits separation of the polysaccharides (eluted in the void volume) from oligosaccharides eluted within the fractionation range of the column (1–10). The concentration of the sample to be loaded is commonly 1–2 mg/ml and should never exceed 1% w/v of the total solutes (Fry 1988). Samples have been eluted in distilled water only (Labrador and Nicolas 1982; Dopico et al. 1986; Dopico et al. 1989) or in a solution containing 20–50 mM NaCl (Huber and Nevins 1979; Huber and Nevins 1981a), as well as in different buffers such as 10 mM Na-phosphate (Revilla et al. 1985), Na-citrate phosphate (Acebes and Zarra 1993) and 50 mM Na-acetate (Gallego and Zarra 1995). However, oligouronides are ionically excluded when chromatographed on Bio Gel P-2 eluted with distilled water (Thibault 1980). Similar results were reported using other columns, e.g. Sepharose CL-4B (Sasaki and Nagahashi 1989).

6.4 Ion Exchange Chromatography

Ion exchange chromatography involves the separation of polysaccharides on the basis of charge. It is useful to obtain fractions enriched in pectins (acid polysaccharides) and hemicelluloses (neutral), e.g. in chick-pea (Revilla et al. 1985) and tomato (Huber and Lee 1988).

7 Conclusions

In vitro autolysis is a useful procedure, allowing the study of structural and compositional changes in the cell wall mediated by wall-associated enzymes. Autolysis can also be used to provide a good indirect estimation of in vivo processes in plant growth and fruit ripening.

However, the use of in adequate buffer, pH, ionic strength or methods to inactivate cell walls (used as controls) must be taken in account, since these factors may seriously falsify the results. Likewise, efforts must be made to characterize cell wall residues on both active and inactive walls after in vitro autolysis assays.

Acknowledgements. We would like to thank Prof. I. Zarra, Dr. E.P. Lorences and Dr. G. Revilla for their critical review of this manuscript. We also thank Mr. D. Lindsey for correcting the English version. Work cited by the author on cell wall autolysis in kiwifruit was financially supported by a DIGICYT grant (PB-89-0554) from the Spanish Government to Prof. I. Zarra. The author is grateful to the Spanish Ministry of Education and Science and the University of Vigo for the fellowships to visit Prof. D. Huber's laboratory at the University of Florida where tomato research was carried out.

References

Acebes JL, Zarra I (1993) Glycanases associated with cell walls of *Cicer arietinum* L.: arabinogalactan degradation. J Exp Bot 44:1089–1094

Albersheim P, Nevins DJ, English PD (1967) A method for the analysis of sugars in plant cell wall polysaccharides by gas-liquid chromatography. Carbohydr Res 5:340–345

Ali ZM, Brady CJ (1982) Purification and characterisation of the polygalacturonases of tomato fruits. Aust J Plant Physiol 9:155–169

Araki N, Takeda H (1992) Purification and properties of a lytic enzyme from the cell wall of *Chlorella ellipsoidea* C-87. Physiol Plant 85:710–718

Avigad G (1975) Colorimetric ultramicroassay for reducing sugars. Methods Enzymol 41:29–31

Blumenkrantz N, Asboe-Hansen G (1973) New method for quantitative determination of uronic acids. Anal Biochem 54:484–489

Bronnenmeier K, Standenbauer WL (1988) Purification and properties of an extracellular β-glucosidase from the cellulitic thermophile *Clostridium steridium*. Appl Microbiol Biotechnol 28:380–386

Bucheli P, Buchala AJ, Meier H (1987) Autolysis in vitro of cotton (*Gossypium hirsutum*) fibre cell walls. Physiol Plant 70:633–638

Buescher RW, Hobson GE (1982) Role of calcium and chelation agents in regulating the degradation of tomato fruit tissue by polygalacturonase. J Food Biochem 6:147–160

Carr DJ, Ng EK (1959) The sequestration of calcium from preparations of wheat coleoptile cell walls. Physiol Plant 12:264–274

Dische Z (1962) Colour reaction of carbohydrates. In: Whistler RL, Wolfrom ML (eds) Methods in carbohydrate chemistry, vol 1. Academic Press, New York, pp 475–514

Dopico B, Labrador E, Nicolas G (1986) Characterisation and localisation of the cell wall autolysis substrate in *Pisum sativum* epicotyls. Plant Sci 44:155–161

Dopico B, Nicolas G, Labrador E (1989) Partial purification of cell wall β-galactosidases from *Cicer arietinum* epicotyls. Relationship with cell wall autolysis processes. Physiol Plant 75:458–464

Dopico B, Nicolas G, Labrador E (1991) Cell wall structure regulates the autolytic process throughout growth of *Cicer arietinum* epicotyls. Physiol Plant 83:659–663

Dubois M, Gilles KA, Hamilton JK, Rebers PA, Smith F (1956) Colorimetric method for determination of sugars and related substances. Anal Chem 28:350–356

Fry SC (1988) The growing plant cell wall: chemical and metabolic analysis. Longman Scientific & Technical, Essex

Galambos JT (1967) The reaction of carbazole with carbohydrates. Anal Biochem 19:119–132

Gallego PP (1992) Papel de la pared celular durante el crecimiento del fruto de kiwi (*Actinidia deliciosa*). Thesis, Universidad de Santiago, Santiago de Compostela, Spain

Gallego PP, Zarra I (1995) Cell wall autolysis in kiwifruit. Changes during fruit development. (submitted)

Gallego PP, Huber DJ, Zarra I (1991) Cell wall hydrolysis in tomato locule gel formation. American Society of Horticultural Sciences, Pennsylvania

Gallego PP, Queijeiro E, Zarra I (1992) Cell wall autolysis in kiwifruit. In: Sassen, MMA., Derksen, JWN, Emons, AMC., Wolters-Arts AMC (eds) 6th Cell wall meeting, Nijmegen, Netherlands, p 161

Geigy Scientific Tables (1984) In: Lentner C (ed) Physical chemistry composition of blood hematology, somatomatric data (Series), vol 3, Ciba-Geigy Ltd, Basel, Switzerland, pp 58–60

Gillet C, Cambier P, Liners F (1992) Release of small polyuronides from *Nitella* cell walls during ionic exchange. Plant Physiol 100:846–852

Goldberg R (1984) Changes in the properties of cell wall pectin methylesterase along the *Vigna radiata* hypocotyl. Physiol Plant 61:58–63

Gross KC (1984) Fractionation and partial characterisation of cell walls from normal and non-ripening mutant tomato fruit ripening. Planta 154:174–179

Gross KC, Wallner SJ (1979) Degradation of cell wall polysaccharides during tomato fruit ripening. Plant Physiol 63:117–120

Hohl M, Hong YN, Schopfer P (1991) Acid- and enzyme-mediated solubilisation of cell-wall β-1.3, β-1.4-D-glucan in maize coleoptiles. Plant Physiol 95:1012–1018

Hobson GE (1964) Polygalacturonase in normal and abnormal tomato fruit. Biochem J 92:324–332

Honda S, Nishimura Y, Takahashi M, Criba H, Kakehi K (1982) A manual method for the spectrophotometric determination of reducing carbohydrates with 2-cyanoacetamide. Anal Biochem 119:194–199

Hoson T (1990) Effect of auxin on autolysis of cell walls in azuki bean epicotyls. Plant Cell Physiol 31:281–287

Huber DJ (1991) Acidified phenol alters tomato cell wall pectin solubility and calcium content. Phytochemistry 30:2523–2527

Huber DJ (1992) The inactivation of pectin depolymerase associated with isolated tomato fruit cell wall: implications for the analysis of pectin solubility and molecular weight. Physiol Plant 86:25–32

Huber DJ, Lee JH (1988) Uronic acid products release from enzymically active cell wall from tomato fruit and its dependency on enzyme quantity and distribution. Plant Physiol 87:592–597

Huber DJ, Nevins DJ (1979) Autolysis of cell wall β-D-glucan in corn coleoptiles. Plant Cell Physiol 20:201–212

Huber DJ, Nevins DJ (1981a) Wall-protein antibodies as inhibitors of growth and of autolytic reactions of isolated cell wall. Physiol Plant 53:533–539

Huber DJ, Nevins DJ (1981b) Partial purification of endo- and exo-β-glucanase enzymes from *Zea mays* L. seedlings and their involvement in cell-wall autohydrolysis. Planta 151: 206–214

Jackman RL, Gibson HJ, Stanley DW (1992) Effects of chilling on tomato fruit texture. Physiol Plant 86:600–608

Jarvis MC (1982) The properties of calcium-bound pectin in plant cell walls. Planta 154:344–346

Keijbets MJH, Pilnik W, Vaal JFA (1976) Model studies on behaviour of pectin substances in the potato cell wall during boiling. Potato Res 19:289–303

Kiss J (1974) β-Eliminative degradation of carbohydrates containing uronic acids residue. Adv Carbohydr Chem 29:299–303

Knee M (1973) Polysaccharide change in cell walls of ripening apples. Phytochemistry 12:1543–1549

Koch JL, Nevins DJ (1989) Tomato fruit cell wall. I. The use of purified tomato polygalacturonase and pectinmethylesterase to identify developmental changes in pectins. Plant Physiol 91:816–822

Labrador E, Nevins DJ (1989) Selected cell wall proteins from *Zea mays*: assessment of their role in wall hydrolysis. Physiol Plant 77:487–492

Labrador E, Nicolas G (1982) Autolytic activities of the cell wall in rice coleoptiles. Effects of nojirimycin. Physiol Plant 55:345–350

Labrador E, Nicolas G (1985) Autolytic of cell walls in pea epicotyls during growth. Enzymatic activities involved. Physiol Plant 64:541–546

Lee S, Kivilaan A, Bandurski RS (1967) In vitro autolysis of plant cell walls. Plant Physiol 42:968–972

Llamazares J, Acebes JL, Zarra I (1987) Cell wall autolysis in *Pinus pinaster* Aiton hypocotyls. Enzymatic activities involved. J Plant Physiol 127:11–22

Loos E, Meindl D (1985) Cell-wall-bound-lytic activity in *Chlorella fusca*: function and characterisation of an endo-mannanase. Planta 166:557–562
Milner Y, Avigad G (1967) A copper reagent for the determination of hexuronic acids and certain ketohexoses. Carbohydr Res 4:359–361
Moustacas AM, Nari J, Diamantidis G, Noat G, Crasnier M, Borel M, Ricard J (1986) Electrostatic effects and the dynamics of enzyme reactions at the surface of plant cells. 2. The role of pectin methyl esterase in the modulation of electrostatic effects in soybean cell walls. Eur J Biochem 155:191–197
Muñoz FJ, Labrador E, Dopico B (1993a) Effect of osmotic stress on the growth of epicotyls of *Cicer arietinum* in relation to changes in the autolytic process and glycanhydrolytic cell wall enzymes. Physiol Plant 87:544–551
Muñoz FJ, Dopico B, Labrador E (1993b) Effect of osmotic stress on the growth of epicotyls of *Cicer arietinum* in relation to changes in cell wall composition. Physiol Plant 87:552–560
Nari J, Noat G, Diamantidis G, Woudstra M, Ricard J (1986) Electrostatic effects and the dynamics of enzyme reactions at the surface of plants cells. 3. Interplay between limited cell-wall autolysis, pectin methyl esterase activity and electrostatic effects in soybean cell walls. Eur J Biochem 155:199–202
Nock LP, Smith C (1987) Identification of polysaccharide hydrolases involved in autolysis degradation of *Zea* cell walls. Plant Physiol 84:1044–1050
Ohno N, Nashimoto Y, Yadomae T (1986) Purification and properties of an exo-(1–3)-β-glucanase from the culture filtrate of *Mucor hiemalis*. Carbohydr Res 158:217–226
Peña MJ, Zarra I, Revilla G (1993) Un problema frecuente: necesidad de encontrar un control adecuado para los estudios de autolisis. In: Aparicio PM (ed) 10th Reunion nacional de la SEFV, 21–24 Sept 1993, Pamplona, Spain, p 318
Pressey R (1986) Extraction and assay of tomato polygalacturonase. HortScience 21:490–492
Pressey R, Avants JK (1973) Two forms of polygalacturonase in tomatoes. Biochim Biophys Acta 309:363–369
Pressey R, Avants JK (1982) Solubilisation of cell walls by tomato polygalacturonases: effects of pectinesterases. J Food Biochem 6:57–74
Revilla G, Sierra MV, Zarra I (1985) Cell wall autolysis in *Cicer arietinum* L. epicotyls. J Plant Physiol 122:147–157
Rushing JW, Huber DJ (1984) In vitro characterisation of tomato fruit softening. The use of enzymatically active cell walls. Plant Physiol 75:891–894
Rushing JW, Huber DJ (1987) Effect of NaCl, pH and Ca^{2+} on autolysis of isolated tomato fruit cell walls. Plant Physiol 7:78–84
Sasaki K, Nagahashi G (1989) Autolysis-like release of pectic polysaccharides from regions of cell walls other than the middle lamella. Plant Cell Physiol 30:1159–1169
Sasaki K, Nagahashi G, Barnet P, Doner L (1987) Autolysis of potato tuber cell walls. In: Cosgrove DN, Knievel DP (eds) Physiology of cell expansion during plant growth. Am Soc Plant Physiol, Rockville, MD, pp 267–269
Satoh H, Takeda H (1989) Detection and first characterisation of a cell-wall lytic activity in *Chlorella ellipsoidea* C-27. Physiol Plant 77:20–26
Schlösser UG (1981) Algal wall-degrading enzymes-autolysines. In: Tanner W, Loewus FA (eds) Encyclopaedia of plant physiology. New Series. Springer, Berlin Heidelberg New York, p 333–351
Schopfer P (1989) pH-dependence of extension growth in *Avena* coleoptiles and its implications for the mechanism of auxin action. Plant Physiol 90:202–207
Scott TA, Melvin EH (1953) Determination of dextran with anthrone. Anal Chem 25:1656–1661
Seara J, Nicolas G, Labrador E (1988) Autolysis of cell wall. Its possible role in endogenous and IAA-induced growth in epicotyls of *Cicer arietinum*. Physiol Plant 72:769–774
Selvendran RR (1975) Analysis of cell wall material from plant tissues: extraction and purification. Phytochemistry 14:1011–1017
Selvendran RR, Stevens BJH, O'Neill MA (1985) Developments in the isolation and analysis of cell walls from edible plants. In: Brett CT, Hillman JR (eds) Biochemistry of plant cell walls. Cambridge Univ Press, Cambridge, pp 39–78
Seymour GB, Harding SE, Taylor AJ, Hobson GE, Tucker GA (1987a) Polyuronide solubilization during ripening of normal and mutant tomato fruit. Phytochemistry 26:1871–1875

Seymour GB, Lasslett Y, Tucker GA (1987b) Differential effects of pectolytic enzymes on tomato polyuronides in vivo and in vitro. Phytochemistry 26:3137–3139
Shiraishi A, Fujii H (1987) Purification of the enzyme that lyases *Sporobolomyces ruberrinus* cell walls and restoration by cysteine of activity inhibited by PCMB. Agric Biol Chem 51:2815–2817
Somogyi M (1952) Note on sugar determinations. J Biol Chem 195:19–23
Taiz L (1984) Plant cell expansion: regulation of cell wall-mechanical properties. Annu Rev Plant Physiol 35:585–657
Thibault JF (1980) Separation of α-D-galacturonic acid oligomers by chromatography on polyacrylamide gel. J Chromatogr 194:315–322
Walter JK, Aach HG (1987) Isolation and characterisation of the enzymes involved in disintegration of the cell wall of *Chlorella fusca*. Physiol Plant 70:485–490

Cell Wall Porosity and Its Determination

Stephen M. Read and Antony Bacic

1 Introduction

Cells must communicate with each other and their environment in order to survive. All higher plant cells are encased in a wall that provides mechanical support to the plant and resists the outward force of turgor pressure exerted by the protoplast. The physicochemical properties of the wall also cause it to act as a molecular and ionic filter, and walls allow passage of some components diffusing from adjacent cells and restrict movement of others. In this review we will present the different methods and approaches that have been used to determine wall porosity, indicate the assumptions inherent in each, and attempt to relate our understanding of wall porosity to the microscopic properties of plant cell walls.

1.1 Biological Significance of Wall Porosity

The primary walls of plant cells are formed of several separate but co-extensive polymer networks (Bacic et al. 1988; Carpita and Gibeaut 1993). Walls of dicots (dicotyledonous plants) and non-graminaceous monocots (type I walls) typically contain interweaving networks of cellulose/xyloglucan, pectin and structural proteins such as extensin. Walls of graminaceous monocots (type II walls) differ in containing a network of cellulose/glucuronoarabinoxylan, embedded in mixed-linkage glucans in expanding cells, but with much smaller amounts of xyloglucan, pectin and extensin-like proteins. The networks are linked by hydrogen bonds and possibly occasional covalent bonds. Deposition of thicker, secondary walls occurs only after cell expansion has ceased.

Primary walls are usually sufficiently permeable to allow the movement of water and solutes of low M_r (relative molecular mass) from the outside of the cell to the plasma membrane, or past cells and through tissues within the apoplast (the connected wall space of adjacent cells). The diffusivity of sucrose in cell walls has been measured at 15–20% of its value in water (Richter and Ehwald 1983). The porosity of walls does, however, restrict the nature and size of macromolecules that can diffuse through the wall, and the size of this filter is important in several aspects of cell physiology.

Many enzymes are secreted into or through the cell wall (Lamport and Catt 1981; Fry 1988), including enzymes that process incoming nutrients (such as

Modern Methods of Plant Analysis, Vol. 17
Plant Cell Wall Analysis
Edited by H.-F. Linskens and J.F. Jackson

invertase), modify wall structure (endohydrolases and transglycosylases during extension growth, peroxidase and pectin methylesterase during wall maturation, and cellulase during abscission), or detect the presence of pathogens (glucanases and chitinases). The access of these enzymes to their substrates in the wall compartment depends on the porosity of the wall.

Plant cells also secrete large polysaccharides and AGPs (arabinogalactan proteins), a process best characterized in suspension-cultured cells (Stevenson et al. 1986) or specialized mucilage-secreting tissues such as root caps or stigmatic papillae (Clarke et al. 1980; Rougier 1981; Fahn 1988). It is unlikely that these hydrophilic molecules, with M_r values between 10^5 and 10^8, would be able to diffuse through the matrix of the wall, which suggests that specific wall pores or other mechanisms of secretion may exist in secretory cells.

In addition, informational molecules are important in intercellular signalling during plant development, and generally need to diffuse through cell walls to wall-bound or plasma-membrane receptors. These molecules include oligosaccharide elicitors, and the proteins controlling embryogenesis in cultured cells or reproductive self-incompatibility in flowering plants. Wall porosity will limit the size of the signalling molecules that can communicate between cells in the plant body.

In many respects, therefore, the interaction of an expanding plant cell with its local environment depends on the functional porosity of its bounding primary wall. These interactions will change during cell growth and differentiation, as will the nature and relative amounts of structural polymers deposited in the cell wall; different cell types and cells in different physiological states can thus be expected to possess walls of different porosities. In addition, it is possible that plant cells can control the extent of their interaction with their environment by modulating wall properties: for example, many cellular responses to wounding or pathogen invasion involve alterations to wall structure that are presumed to reduce the susceptibility of walls to digestion or penetration (Aist 1983; Bradley et al. 1992), and that may also reduce wall porosity.

1.2 Assumptions and Definitions

Pore size is derived from functional assays as the diameter of the largest tracer molecule that can pass through walls in a given time; the diameter of a molecule is twice its Stokes' radius calculated from physical measurements (Squire 1981; Peters 1986). However, the relationship between molecular diameter and M_r is not the same for poly(ethylene glycol)s (PEGs), dextrans and globular proteins, the three types of molecules generally used as porosity tracers (Fig. 1). These molecules have different tertiary structures and occupy different hydrodynamic volumes in aqueous solution: PEGs have a relatively open conformation, dextrans form a random coil, and proteins are more tightly folded. Thus, PEG-4000 (M_r 3000–3700), dextrans of M_r 10 000, and actin and ovalbumin (both of M_r 43 000) all have molecular diameters of 4.5–4.7 nm (Carpita et al. 1979; Squire 1981; Peters

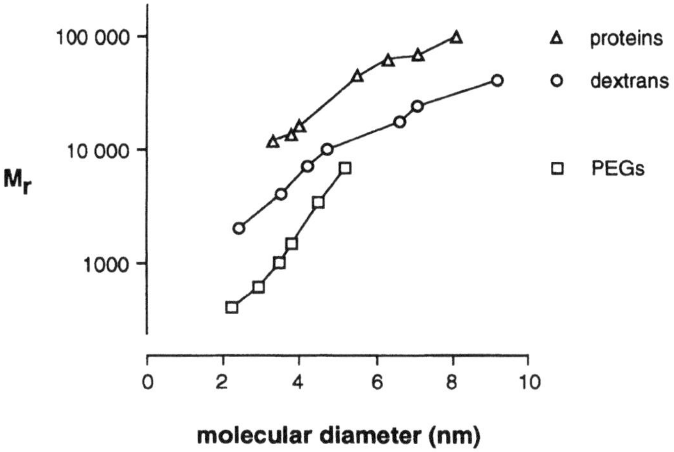

Fig. 1. Relationship between molecular diameter and M_r for PEGs, dextrans and globular proteins. Data are taken from Carpita et al. (1979), Peters (1986) and references therein, and are plotted for PEG-400. -600, -1000, -1540, -4000 and -6000, for various dextrans, and for cytochrome c, RNAse A, myoglobin, ovalbumin, haemoglobin, BSA and the $F(ab)_2$-fragment of IgG

1986). Within each series, M_r varies with the cube of molecular diameter, and small variations in pore size can thus cause large changes in the M_r of molecules that can penetrate walls.

Treating molecules as simple spheres of fixed size can also on occasion be misleading. Some proteins, such as IgG (immunoglobulin G), are ellipsoidal, with at least one short dimension (Peters 1986), and may penetrate the wall to a greater extent than globular proteins of the same M_r, as long, thin molecules can diffuse slowly through pores that are only slightly larger than their short axis (Meredith and Nathans 1982). Furthermore, the solution concentration of molecules can influence their conformation: for example, PEG molecules are elongate and relatively rigid at lower concentrations, but become more disordered and spherical as their concentration is increased (Michel and Kaufmann 1973). Chemical modification, for example by attachment of fluorescent reporter groups, may also alter molecular size or cause aggregation. Charged molecules will be repelled from pores bearing fixed charges of the same sign, and thus have a reduced rate of diffusion through the wall. The effective porosity of walls thus results from interactions between the physicochemical properties of the solute and the chemical and physical environment in the wall.

We distinguish here three types of technique for determining wall porosity. *Ultrastructural methods* analyse isolated cell walls, for example by EM (electron microscopy). *Bulk exclusion techniques* include solute exclusion, which was the basis of the seminal paper on determination of wall porosity (Carpita et al. 1979); solute exclusion is an osmotic method that determines the size of solutes that can penetrate into the wall. *Functional assays* detect the transport of appropriately labelled tracer molecules across an intact wall, and have the advantage of being performed on whole cells under close to physiological conditions; they also assay more directly transport across the entire breadth and depth of the wall, but do not give data on variations in pore diameter through the wall, or determine whether pores are specific, fixed channels or arise from the random spatial arrangement of wall polymers. We will use the categories of tracer movement defined by Baron-Epel et al. (1988), as they are easy to visualize: *unhindered movement* (passing through the wall in less than 10 min), *hindered movement* (needing 1–2.5 h to pass through the wall), and *excluded* (not passing through the wall at all).

2 Microscopic Visualization of Wall Pores

Electron microscopy (EM) offers the possibility of observing wall pores directly, but individual pores have only been reported for a limited number of cell types; references to early work are given in Carpita et al. (1979) and Carpita (1982). The stigmatic papillar cells of *Gladiolus* have a thin wall overlaid by a cuticle, and are specialized for secretion of large amounts of mucilage. Clarke et al. (1980) showed by freeze-fracture techniques that the cellulosic layers of these walls were penetrated by a system of channels approximately 50 nm in diameter, which were suggested to be the route taken by secreted material passing through the wall. An extensive network of similar wall channels was detected in freeze-substituted trichomes from leaves of the bromeliad *Brocchinia* (Owen and Thomson 1991); these cells are specialized for uptake of macromolecular nutrients. The possibility that these holes or channels could be produced artefactually, or else enlarged, during processing of tissue for microscopy should be kept in mind. In addition, observed structures may be blocked with material being secreted rather than representing empty channels, or may be the remnants of plasmodesmata (Robards and Lucas 1990) or plasma tubules (Harris and Chaffey 1986).

However, apart from these and similar examples of cells specialized for secretion and cells specialized for uptake of nutrients, walls of most cells do not appear to contain specifically constructed pore structures. Observation of shadowed replicas of rapidly frozen, deep-etched primary cell walls from onion (*Allium*) parenchyma visualized pores as random gaps in the network of wall polymers rather than as specific structures (McCann et al. 1990). The pores were mostly 10 nm in diameter, but would have been artefactually enlarged during the deep-etching process. Treatment of walls with cold dilute alkali, which dissolves primarily pectin, gave a more open-textured wall with a significant increase in porosity, but otherwise caused little change in the large-scale structure of the wall because other polymer networks remained. The porosity of these primary walls thus appeared to be determined by pectin, as this formed the polymer network with the smallest pores.

3 Bulk Exclusion Techniques

3.1 Solute Exclusion

Carpita et al. (1979) applied the osmotic technique of "solute exclusion" to determine the pore size of walls of living plant cells. The cells or tissue under test are immersed in hypertonic solutions of solutes of different molecular sizes, such as PEGs or dextrans, and may undergo either plasmolysis or cytorrhysis (Fig. 2). *Plasmolysis*, where the protoplast shrinks and pulls away from the inner surface of the cell wall, implies that the solute has been able to penetrate the wall and accumulate in the periplastic space around the protoplast. By contrast, *cytorrhysis*, where both the protoplast and wall shrink and the whole cell collapses, implies that the solute cannot penetrate the wall. The size of the largest molecules causing plasmolysis is thus a measure of the diameter of cell wall pores. The rapidity of this technique means that only the overall porosity of the wall matrix is measured, and that specialized secretory channels do not affect the results.

Carpita et al. (1979) applied solute exclusion to five different cell types, from cotton fibres to suspension-cultured cells, and found remarkable consistency in the results, with limiting pore diameters between 3.5 and 5.2 nm (see Table 1 for a compilation of results from this and other techniques). For example, root hairs of radish (*Raphanus*) were plasmolysed by a PEG-600 solution (molecular diameter 2.9 nm), exhibited both plasmolysis and cytorrhysis in a PEG-1000 solution (3.5 nm), and showed only cytorrhysis in a PEG-1540 solution (3.8 nm), which gave a limiting diameter of 3.5 nm. Hoggart and Clarke (1984) subsequently used the technique on stigmatic papillae and pollen tubes of *Gladiolus* (Fig. 2), and found similar values (3.4–4.5 nm), but solute exclusion does not appear to have been greatly used since.

The solute concentrations or water potentials of the solutions used were not quoted by Carpita et al. (1979), but Hoggart and Clarke (1984) used PEG solutions at 40–80% (w/v). These PEG solutions all have substantial negative "matrix potentials" (Mexal and Reid 1972; Mexal et al. 1975) in addition to their negative water potentials due to colligative properties, and it is preferable to use a series of solutions of equivalent total water potential to give consistent rates and extents of extraction of water (Money 1980). Larger solutes that cause plasmolysis often first cause cytorrhysis, showing that water leaves the wall compartment faster than these solutes enter, but this kinetic dehydration appears to be readily reversible (Carpita 1982). Tepfer and Taylor (1981) suggested that very high solute concentrations could also dehydrate the cell wall permanently, with removal of bound water altering wall structure and significantly reducing porosity; some shrinkage of walls was detected in concentrated solutions of PEG-6000 or BSA. A correction for wall shrinkage can be easily applied (Money 1980). The mean size of PEG molecules is significantly reduced in concentrated solution (Michel and Kaufmann 1973), and using solutions of equivalent water potential would also reduce any effect of changes in molecular shape. It is also possible that high concentrations of

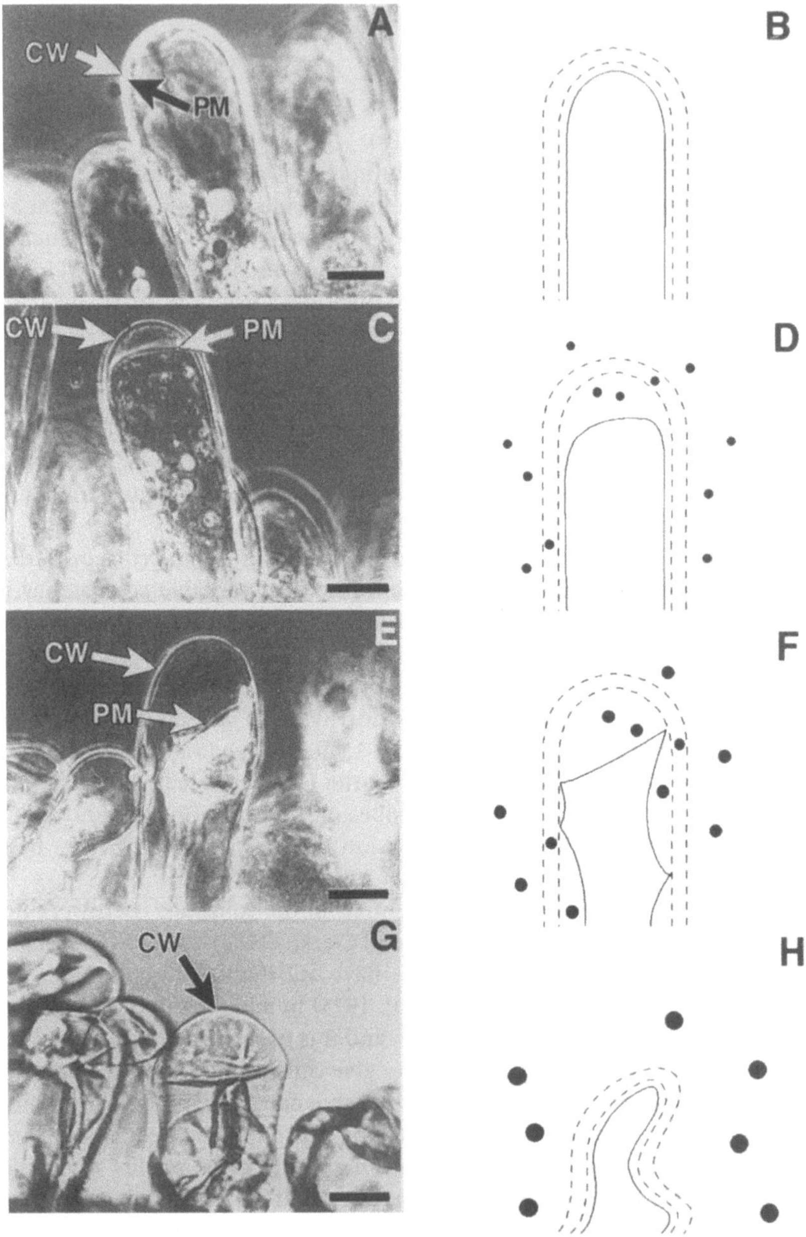

Fig. 2A–H. Plasmolysis and cytorrhysis induced in stigmatic papillae of *Gladiolus*. **A** In isotonic solution; **C** slightly plasmolyzed in 0.4 M mannitol (7.3% w/v); **E** strongly plasmolyzed in 0.4 M PEG-1000 (40% w/v); **G** exhibiting cytorrhysis in 0.2 M PEG-4000 (80% w/v). *CW* Cell wall; *PM* plasma membrane. Adapted from Hoggart and Clarke (1984). The diagrams in **B**, **D**, **F** and **H** show the behaviour of the cell wall (*dotted lines*) and plasma membrane (*solid line*) in the presence of these different solutes of varying size (*black dots*). **B** Control; **D** slightly plasmolyzed; **F** strongly plasmolyzed; **H** exhibiting cytorrhysis

Table 1. Limiting diameters measured for pores in plant cell walls

Cell type	Plant	Pore diameter (nm)	Method	References
Bulk exclusion techniques				
Root hairs	*Raphanus*	3.5–3.8	Solute exclusion	Carpita et al. (1979)
Ovule fibres	*Gossypium*	3.5–3.8	Solute exclusion	Carpita et al. (1979)
Suspension cells	*Acer*	3.8–4.0	Solute exclusion	Carpita et al. (1979)
Leaf parenchyma	*Xanthium*	4.5–5.2	Solute exclusion	Carpita et al. (1979)
Leaf parenchyma	*Commelina*	4.5–5.2	Solute exclusion	Carpita et al. (1979)
Pollen tubes	*Gladiolus*	3.5–4.5	Solute exclusion	Hoggart and Clarke (1984)
Stigmatic papillae	*Gladiolus*	3.5–4.5	Solute exclusion	Hoggart and Clarke (1984)
Suspension cells	*Chenopodium*	4.5–5.2	Exclusion of PEGs	Gogarten (1988)
Hypocotyls	*Phaseolus*	5.5	Gel filtration	Tepfer and Taylor (1981)
Suspension cells	*Chenopodium*	4.5–5.2	Gel filtration	Gogarten (1988)
Suspension cells	*Chenopodium*	5.3–5.6	Gel filtration	Ehwald et al. (1991, 1992)
Tracer uptake techniques				
Cultured roots	*Glycine*	9.2	Uptake of F-dextran	Baron-Epel et al. (1988)
Leaf trichomes	*Brocchinia*	6.6	Uptake of F-dextran	Owen and Thomson (1991)
Suspension cells	*Nicotiana*	8.6	Uptake of F-dextran	Shedledzky et al. (1992)
Suspension cells	*Hordeum*	8.6	Uptake of F-dextran	Shedledzky et al. (1992)
Suspension cells	*Morinda*	5.7	Uptake of F-dextran	O'Driscoll et al. (1993)
Pollen tubes	*Nicotiana*	4.5	Uptake of F-dextran	O'Driscoll et al. (1993)
Pollen tubes	*Tradescantia*	6.6	Uptake of F-dextran	O'Driscoll et al. (1993)
Cultured roots	*Glycine*	7.9	Uptake of proteins	Baron-Epel et al. (1988)
Suspension cells	*Glycine*	7.2	Uptake of proteins	Horn et al. (1992)
Suspension cells	*Morinda*	5.7	Uptake of gold sol	O'Driscoll et al. (1993)

PEG could block the pores by precipitating any soluble polymers being secreted through the wall (Iraki et al. 1989). Taken together, these considerations suggest that solute exclusion may underestimate the size of wall pores.

3.2 Long-Term Exclusion of PEG (Polyethylenglycol)

Measurement of the exclusion of a range of PEGs from walls of growing cells can also be performed at lower solute concentrations and over longer times than with the solute exclusion technique. Suspension-cultured cells of tomato (*Lycopersicon*) were shown to take up [^{3}H]PEG-4000 into only one compartment, from which the PEG could readily be eluted by washing in fresh medium, and this was concluded to be the wall space (Handa et al. 1982). Gogarten (1988) then cultured cells of *Chenopodium* for 35 days in media containing PEGs at 4% (w/v) and determined the refractive index of the conditioned media. All PEGs were concentrated in the medium due to cell growth, but larger PEGs were additionally concentrated due to their exclusion from the walls: PEG-1540 (3.8 nm) was almost fully included in the wall compartment, PEG-4000 (4.5 nm) gave intermediate results, and PEG-6000 (5.2 nm) was excluded.

3.3 Gel Filtration

Tepfer and Taylor (1981) prepared cell walls from bean (*Phaseolus*) hypocotyls by grinding frozen tissue, followed by sonication in buffer and centrifugation. The walls were packed into a column and used as a gel-filtration matrix for chromatography of a range of proteins of different sizes; 0.5 M NaCl was added to the elution buffer to block ionic interactions between the walls and charged protein molecules. Cytochrome c, RNAse A and myoglobin (4.0 nm or below) were almost fully included in the matrix, ovalbumin (5.5 nm) was substantially excluded, and haemoglobin and BSA (6.3 nm and over) were almost fully excluded. Gogarten (1988) packed a gel-filtration column with living suspension-cultured cells of *Chenopodium*, reasoning that isolation and fragmentation of walls could alter their microstructure. Different PEGs were dissolved in medium at sufficiently low concentration (less than 2% w/v) to have no effect on wall structure, and applied to the column. PEG-4000 (4.5 nm) was partly excluded from the walls, eluting between fully included molecules (such as PEG-1540, 3.8 nm) and fully excluded molecules (PEG-6000, 5.2 nm).

The interpretation of these results depends on how much of the wall matrix must be accessible to a given molecule for detectable amounts to penetrate completely through the wall (Tepfer and Taylor 1981; Carpita 1982). Following the classification of Baron-Epel et al. (1988), molecules that chromatograph close to the fully included volume, and that can penetrate the majority of the wall space, would be almost unhindered in their movement through the wall; the value of 4.0–4.5 nm measured this way is in accordance with the overall porosity determined by solute exclusion (Carpita et al. 1979; Carpita 1982). Larger molecules, up to 6.3 nm, can penetrate only a certain portion of the wall space and will diffuse to the plasma membrane at considerably reduced rates.

The included volume on the whole-cell columns of Gogarten (1988) was substantially larger than the volume of walls determined by electron microscopy, probably because spaces at cell corners and inside dead cells were only accessible after diffusion through a cell wall. This principle also applies to columns packed with intact but empty cells treated with organic solvents and protease ("microcapsules"), which separate solutes into two fractions, one fully included and one fully excluded (Ehwald et al. 1991, 1992). The size limit of permeation (the maximum diameter of a molecule capable of equilibrating with the intracellular space) was determined with PEGs, dextrans and neutral proteins to be 5.3–5.6 nm, and acidic proteins were excluded to a greater extent than were neutral proteins, consistent with the limiting polymer being negatively charged. The permeation limit increased progressively to over 12 nm with treatments that de-esterified, depolymerized or extracted pectin from the walls (Ehwald et al. 1992).

4 Tracer Techniques: Uptake of Molecules or Particles

Observation of the ability of living cells to take up molecules or colloidal particles of known size provides a direct functional assay of wall porosity. Molecules diffusing through the wall have a range of possible fates, including accumulation in the periplastic space between the inner face of the wall and the plasma membrane, transport across the plasma membrane into the cytoplasm, or endocytosis or uptake into vesicular compartments inside the cell. A note of caution thus applies because, depending upon the particular assay used for movement of the tracer, steps other than diffusion across the wall can be rate-limiting.

4.1 Small Molecules

The apoplast is generally permeable to water, salts and nutrients, and the uptake of small molecules into cells has only been used to assay wall properties in cases where the wall is considered to be especially impermeable. The best-studied examples of altered porosity arise in specialized cells, such as those with walls containing callose or hydrophobic polymers.

Cells with Callosic Walls. Callose, a (1–3)-β-linked glucan, is deposited as an inner wall layer at a variety of sites during plant development, and is presumed to have a sealing function (Currier 1957; Evert and Derr 1964). For example, a thick callosic wall is deposited around pollen mother cells (meiocytes) at meiotic prophase in developing anthers, and remains around the resultant tetrad of haploid microspores, temporarily isolating the initial cells of the male gametophytic generation from sporophytic tissue (Mascarenhas 1975). The walls are subsequently dissolved to release the microspores.

Heslop-Harrison and MacKenzie (1967) supplied [2-^{14}C]thymidine to *Lilium* buds and inflorescences, and analysed its distribution within anthers by dry-emulsion autoradiography; treatment of preparations with acid allowed soluble tracer to be distinguished from tracer incorporated into DNA. Radioactivity only entered the meiocytes early in meiotic prophase, and did not enter microspores until after their release from the callosic wall. However, exclusion may also result from the cessation of cell growth, DNA synthesis or thymidine transport during the period that the cells are surrounded by callose (Heslop-Harrison and MacKenzie 1967; Mascarenhas 1975). Isolated microspore tetrads of *Lilium* and *Gladiolus* also excluded FDA (fluorescein diacetate), while microspores released from their callose wall, either mechanically or by its normal developmental dissolution, took up FDA and converted it intracellularly to fluorescein (Knox and Heslop-Harrison 1970). The uncharged FDA molecule is able to cross membranes directly, and so these data suggest that either the interface between the plasma membrane and the wall, or the wall itself, become less permeable when a callosic wall is present around the microspores.

On the other hand, a few reports indicate that uptake of an unusual range of small solutes can occur through the callosic walls of developing microspores: these include ferric iron (Rowley and Dunbar 1970), cerous ions (Rodríguez-García and Majewska-Sawka 1992) and substrates for the cytological staining of succinic dehydrogenase and cytochrome oxidase (Sauter and Marquardt 1970). The relative impermeability or decreased porosity of callosic walls has therefore not been conclusively established.

Cells with Suberized Walls. Reduction of the permeability of cell walls to water by deposition of hydrophobic polymers, such as lignin, cutin and suberin, influences the path of water movement through the plant and thus the sites of uptake of dissolved nutrients into the symplast (the connected cytoplasmic compartments of adjacent cells). Localized deposits of suberin are found in regions of cell walls of the endodermis in roots (the Casparian strip) and the bundle sheath cells in leaves (the mestome sheath), places where the transpiration stream is forced to enter the symplast (O'Brien and Carr 1970; Clarkson and Robards 1975; Canny 1990). Suberized regions are generally identified as barriers to the movement of fluorescent dyes or brighteners through the plant (Peterson et al. 1981; Canny 1990). Canny (1988) used this technique to observe 100-nm-wide paths for solute diffusion through the suberized walls. An apoplastic barrier also occurs in the side walls of cells at the base of secretory trichomes (Fahn 1988). Altered cell walls developed in response to pathogen attack also generally contain hydrophobic polymers (Aist 1983).

4.2 Macromolecules

Most work on the movement of macromolecules through cell walls has been performed with homologous series of dextrans. The molecular sizes of these spherical molecules are given in Carpita et al. (1979) and Peters (1986; see also Fig. 1). Dextrans can readily be detected microscopically after derivatization with fluorescein (giving F-dextrans), but individual batches should be tested for the presence of insoluble aggregates or low-M_r fluorescent molecules such as free FITC, and purified if necessary, or erroneous results can be obtained (Preston et al. 1987; Cole et al. 1990). Similarly, control experiments should be performed to check that no degradation of dextrans to smaller molecules is occurring during incubation with cells (O'Driscoll et al. 1993). Detection of the diffusion of F-dextrans across cell walls requires the cells to be plasmolyzed or treated with detergent (Baron-Epel et al. 1988; Shedletzky et al. 1992), otherwise the method also assays transport of the labelled macromolecules across the plasma membrane (Cole et al. 1990; O'Driscoll et al. 1993).

Baron-Epel et al. (1988) measured the rates of movement of F-dextrans through the walls of cultured soybean (*Glycine*) root cells plasmolyzed with 0.5 M mannitol, by examining both the distribution of fluorescence and its redistribution after photobleaching. F-dextrans with M_r values up to 17900 (6.6 nm) showed unhindered movement, passing through the wall in less than 10 min; F-dextran of M_r 41000 (9.2 nm) showed hindered movement, needing 1–2.5 h to pass through the wall; and F-dextrans of M_r over 65200 (11 nm and above) were excluded. The limiting diameter of wall pores was thus determined as about 9 nm. Cole et al. (1990) showed that F-dextran of M_r 20000 (6.6 nm) was taken into the vacuoles of cultured carrot (*Daucus*) cells while F-dextran of M_r 70000 (11 nm) was not, and cultured cells of *Morinda* took up F-dextrans of M_r 9400 (4.5 nm) but failed to take up F-dextrans of M_r 17200 (6.6 nm) (O'Driscoll et al. 1993). In these cases, however, cells were not plasmolyzed, and differences in the rates of fluid-phase endocytosis of the dextrans may have affected the results.

Uptake of F-dextrans has also been used to study the wall porosity of specialized cells. Cultured pollen tubes of *Tradescantia* rapidly took up F-dextran of M_r 9400 (4.5 nm) and slowly took up F-dextran of M_r 17200 (6.6 nm), while cultured pollen tubes of tobacco (*Nicotiana*) were less permeable, slowly taking up F-dextran of M_r 9400 (4.5 nm), but excluding F-dextran of M_r 17200 (6.6 nm) (O'Driscoll et al. 1993). The high level of PEG-6000 in the culture medium for *Nicotiana* pollen tubes may have decreased measured wall porosity. Fluorescence was initially detected at pollen-tube tips, but it was not clear which of the two potentially rate-limiting processes, movement across the wall or endocytosis, was restricted to this region. Chang and Trevithick (1974) have suggested that the newly formed wall at the tips of fungal hyphae contains pores that are larger than those in the majority of the wall, allowing secretion of protein molecules to take place at the growing hyphal tip and a similar situation could occur in pollen tubes. Leaf trichomes of the bromeliad *Brocchinia* also took up F-dextrans of M_r up to 17200 (6.6 nm) (Owen and Thomson 1991).

Proteins of known size can also be used as tracers of wall porosity. Fluorescently labelled wheat-germ agglutinin (5 nm) and bovine serum albumin (7.2 nm) passed through the walls of plasmolyzed cultured soybean cells, while soybean agglutinin (8.6 nm) was excluded (Baron-Epel et al. 1988). Biotinylated proteins entered cultured soybean cells by receptor-mediated endocytosis via the biotin receptor, but the rate of this process decreased with increasing protein size probably because the rate-limiting step was diffusion through the wall (Horn et al. 1992). The rate of uptake was maximal for insulin (2.6 nm), presumably unhindered in diffusion through the wall, while RNAse (3.8 nm), soybean trypsin inhibitor (4.8 nm) and BSA (7.2 nm) were hindered and IgG (11 nm) was fully excluded. Inhibition of auxin-induced extension of maize (*Zea*) coleoptiles by antisera specific for wall proteins or carbohydrates implies that at least under some experimental condition molecules as large as IgG can penetrate cell walls (Hoson and Nevins 1989; Inouhe and Nevins 1991), although the asymmetric nature of the IgG molecule (Peters 1986) must be taken into consideration. The log-linear relationship between M_r and the rate of protein uptake (Horn et al. 1992) is similar to the relationship between M_r and elution position on gel filtration (Tepfer and Taylor 1981), which implies that the rate at which a protein diffuses through the wall to the plasma membrane is related to the proportion of the wall space available to that protein. This is good evidence that pores are randomly and evenly distributed throughout the depth of the wall.

Treatment of cultured soybean cells with a large variety of reagents, including cellulysin, protease and low pH, had no effect on the rates of movement of F-dextrans across the wall of plasmolyzed cells (Baron-Epel et al. 1988). However, exposure to a pectinase preparation significantly enlarged the wall channels, allowing unhindered uptake both of F-dextrans of M_r up to 65 200 (11 nm) and of soybean agglutinin (8.6 nm); F-dextran of M_r 157 000 (18 nm) was still excluded. This experiment provided direct evidence that pectin microstructure determines overall porosity. Horn et al. (1992) also showed that the rates of uptake of smaller proteins could be increased by mild digestion of walls with polygalacturonase, although in this case the upper size limit for penetration was not increased. Meiners et al. (1991) demonstrated that an increase in wall porosity and also permeabilization of the plasma membrane could be reversibly induced by treatment of cells with 0.1 mg ml^{-1} saponin, whereas pectinase increased wall porosity irreversibly but had no effect on the plasma membrane. Saponins are plant glycosides that perturb membranes, but the mechanism by which they increase wall porosity is unclear.

Shedletzky et al. (1992) investigated the uptake of F-dextrans into cultured tobacco (*Nicotiana*) and barley (*Hordeum*) cells; the cells were treated with detergent to ensure that the limiting step was diffusion through the cell wall. The limiting pore diameter for both cell types was calculated as 8.6 nm, similar to the findings on cultured soybean cells (Baron-Epel et al. 1988). The walls of tobacco cells adapted to growth on medium containing DCB (2,6-dichlorobenzonitrile, an inhibitor of cellulose synthesis) lacked a cellulose/xyloglucan network, and the remaining wall consisted mostly of Ca^{2+}-birdged pectins; wall porosity, however, was unaltered. Barley cells adapted to DCB also contained little cellulose in their walls, but showed increased levels of non-cellulosic polysaccharides (mixed-link glucan and glucuronoarabinoxylan) and an altered pattern of cross-linking by phenolic components; wall porosity was reduced, with a limiting pore diameter of about 6.6 nm. These results are consistent with the pectic matrix being the major determinant of porosity in the type I walls of dicot plants such as tobacco, and suggest that wall porosity in the type II walls of graminaceous monocots such as barley is controlled by the degree of cross-linking of non-cellulosic polysaccharides (Shedletzky et al. 1992; Carpita and Gibeaut 1993).

4.3 Particles

Uptake of electron-opaque colloidal sols has been used to measure the capacity of cells for fluid-phase endocytosis (Robinson and Hillmer 1990), and has also given information on wall porosity. O'Driscoll et al. (1993) incubated cultured cells of *Morinda* with a gold sol of particle size range 1.9–7.9 nm, but observed only particles of up to 5.7 nm entered cells, suggesting that the limiting pore diameter was about this value. Most particles were observed inside cells during the log phase of culture growth, although it was not clear whether this represented a maximum wall porosity or a maximum rate of fluid-phase endocytosis.

5 Conclusions

5.1 Integration of Results from Different Methods

Only a small literature describes direct measurements of the porosity of plant cell walls, with little comparison of different techniques on the same plant material. Consensus on the most appropriate techniques has thus not yet been attained. The range of wall porosity values determined from functional assays is given in Table 1; EM data are not tabulated because of the difficulties of allowing for dimensional changes during specimen preparation. Solute exclusion thus gives limiting pore diameters in the range 3.5–5.2 nm, measurement of the long-term exclusion of PEG gives 4.5–5.2 nm, and gel filtration gives 4.5–5.6 nm; these techniques all measure bulk penetration of material into the cell wall. In contrast, techniques that measure the diffusion of trace amounts of dextrans, proteins or colloidal particles to the plasma membrane give, with one exception, distinctly larger limiting pore diameters, in the range 5.7–9.2 nm. While part of this variation undoubtedly represents real differences in the porosity of different cell types, the majority derives from differences in the methods and their inherent assumptions.

The conclusion that techniques assessing bulk movement of solute give a smaller pore size, while methods measuring the penetration of only a small proportion of an external tracer detect larger pores, suggests that there is a continuous range of pore sizes in the wall. Abundant 4–5 nm pores probably contribute to bulk uptake or exclusion, and less frequent 6–9 nm pores allow larger molecules to penetrate more slowly. The concept of a single limiting pore diameter thus does not appear to apply, and there is a significant molecular size range where *hindered movement* (Baron-Epel et al. 1988) can occur; the continuous range of pore size distribution is shown directly by the data of Horn et al. (1992). The similar dependence on M_r of the degree to which proteins penetrate the bulk of the wall matrix, determined by gel filtration (Tepfer and Taylor 1981), and their rate of access to the plasma membrane (Horn et al. 1992), implies that pores are randomly and evenly distributed throughout the depth of the wall.

Taken together, the measured porosity means that globular proteins up to M_r 15–25 000 can diffuse relatively unhindered through primary cell walls, while proteins up to about M_r 75–100 000 diffuse progressively more slowly. These values are significantly larger than those measured for symplastic permeability through plasmodesmata, which allow free movement only of peptides of M_r below 1000 (Robards and Lucas 1990). Although most wall-localized enzymes will exhibit hindered diffusion through the apoplast, these values do not necessarily impose a great limit on apoplastic biochemistry; slow diffusion of proteins, over for example a couple of hours, may be sufficient to mediate many biological processes. Larger enzymes will be confined to the cell wall because of their size, and secretion of large molecules such as the polysaccharides and proteoglycans of gums and mucilages will require special wall modifications and may occur in specialized regions of the cell. Detailed studies of protein mobility within walls are needed in the important region between M_r 25 000 and 75 000, a three-fold difference in M_r which corresponds to only a 45% difference in pore diameter.

5.2 Pore Structure

There is considerable evidence that in dicot cells the network of polymers with the finest mesh size, and which provides the background porosity of the wall, is pectin. Chemical or enzymatic removal of pectin can increase both diffusion rates and the size of molecules that can diffuse through the walls (Baron-Epel et al. 1988; Ehwald et al. 1992; Horn et al. 1992), and the resultant more open mesh of wall fibrils is consistent with the functionally measured increase in porosity (McCann et al. 1990). The component that determines porosity in the walls of graminaceous monocots is less clear, but Shedletzky et al. (1992) suggested that this could be the cross-linked glucuronoarabinoxylan network.

Chemically, pectin is a highly complex acidic polymer that mediates many biological interactions, but little is known about the detailed structure of the pectin gels formed in cell walls (Carpita and Gibeaut 1993); pores presumably result from irregularities in its random spatial arrangement. Pectins are the most mobile components of dicot cell walls (Taylor et al. 1990), and it is possible that the larger pores that allow diffusion of molecules up to 6–9 nm diameter are produced by the transient opening ("breathing") of smaller pores. However, a large fraction of pectin occurs in rigid domains (Taylor et al. 1990), and the degree of formation of junction zones between pectin backbones could also affect wall porosity. In addition, the lining of pores with acidic residues means that molecules negatively charged at the apoplastic pH of 5 will be repelled and have a reduced rate of diffusion through the wall. Changes in wall pH or surface charge may thus directly affect protein mobility and the access of enzymes to wall-bound substrates.

Most techniques measure the overall porosity of the gel matrix of the wall, but cells specialized for secretion or absorption may require an additional mechanism for passage of large molecules through their walls. There is occasional EM evidence for distinct physical channels running between the plasma membrane and the outer surface of the wall, and these apoplastic channels have been detected in both absorptive cells (Owen and Thomson 1991) and secretory cells (Clarke et al. 1980). The overall porosity of *Gladiolus* stigmatic papillar walls was determined by solute exclusion as 3.5 nm (Hoggart and Clarke 1984), and was unrelated to the 50-nm channels observed in restricted sections of these walls by Clarke et al. (1980). Flow of water and solutes must therefore occur more rapidly through the whole wall area than through these channels, possibly because they contribute only a small fraction of the total pore area or are normally blocked by macromolecules being secreted. There is also physiological evidence for secretory pores. Secretion of EPS (extracellular polysaccharide) by cultured sycamore (*Acer*) cells was rapidly and reversibly blocked by addition of EPS at relatively high concentration (1% w/v), without affecting polysaccharide synthesis (Paull and Jones 1978). Added EPS had no effect on the release of polysaccharide from protoplasts, and it was proposed that EPS inhibited secretion by direct interaction with secretory pores in the walls. Suspension-cultured cells of tobacco (*Nicotiana*) adapted to growth on medium containing 30% (w/v) PEG-8000, which does not penetrate the wall, were also inhibited in their secretion of EPS (Iraki et al. 1989). Transfer of cells into medium containing iso-osmotic mannitol caused sudden release of EPS, consistent with the high concentration of PEG-8000 inhibiting secretion by precipitating EPS onto the wall surface or into wall pores.

In some cases, however, secretion of large molecules requires complete degradation or shedding of the wall (Fahn 1988). Germination of barley in the absence of gibberellic acid prevents breakdown of the aleurone cell wall, and enzymes secreted by the aleurone accumulate in the inner region of the wall (Fincher 1989). In the presence of gibberellic acid, movement of enzymes through spaces in the degrading aleurone walls can be followed both cytochemically (Ashford and Jacobsen 1974) and by EM immunocytochemistry (Gubler et al. 1987). Root caps secrete very large slime polysaccharides to ease passage of the growing root through the soil (Rougier 1981). The root cap cell walls act as a barrier to the passage of slime molecules, which accumulate in the periplastic space and are released periodically when root cap cells are shed and their walls disrupted (Paull and Jones 1976). Both these examples thus provide evidence that passage of secreted molecules can be blocked because of the impermeability of the cell wall.

5.3 Variations in Wall Porosity

Localized or short-term changes in wall pH, Ca^{2+} concentration and the degree of carboxyl esterification could provide a dynamic control of wall porosity through affecting the degree of formation of junction zones in the pectic polymers of type I walls. Porosity of the type II walls of grasses could be similarly controlled via the degree of branching and esterification of glucuronoarabinoxylan (Carpita and Gibeaut 1993). These changes will necessarily be restricted by the intercalation of the different wall polymer networks.

Wall composition changes during growth, differentiation and environmental responses (Bacic et al. 1988), as does the rigidity of wall polymers (Wallace et al. 1993). Many responses to fungal or bacterial attack involve rapid formation of altered cell walls, including deposition of callose, suberin, cutin and lignin, structural proteins, and increased cross-linking of existing polymers (Aist 1983; Bacic et al. 1988; Bradley et al. 1992). This presumed toughening or hardening of the cell wall creates a major change in its architecture, and is presumed to create permeability barriers restricting diffusion of enzymes and toxins from the pathogen to healthy cells, and water and nutrient from the host to the pathogen; however, direct evidence for altered porosity or permeability in these cases is lacking.

5.4 Future Developments

Although consensus is emerging on the functional porosity of most plant cell walls, it is not clear whether wall pores result solely from gaps in the polymer lattice or if additional, specifically constructed pores exist in walls of specialized cells. An increased knowledge of the architecture of living walls is thus needed to further our understanding of the molecular nature of wall pores and the mechanisms that regulate wall porosity.

References

Aist JR (1983) Structural responses as resistance mechanisms. In: Bailey JA, Deverall BJ (eds) The dynamics of host defence. Academic Press, Sydney, pp 33–70

Ashford AE, Jacobsen JV (1974) Cytochemical localization of phosphatase in barley aleurone cells: the pathway of gibberellic-acid-induced enzyme release. Planta 120:81–105

Bacic A, Harris PJ, Stone BA (1988) Structure and function of plant cell walls. In: Preiss J (ed) The biochemistry of plants, vol 14. Academic Press, London, pp 297–371

Baron-Epel O, Gharyal PK, Schindler M (1988) Pectins as mediators of wall porosity in soybean cells. Planta 175:389–395

Bradley DJ, Kjellbom P, Lamb CJ (1992) Elicitor- and wound-induced oxidative cross-linking of a proline-rich plant cell wall protein: a novel, rapid defense response. Cell 70:21–30

Canny MJ (1988) Water pathways in wheat leaves. IV. The interpretation of images of a fluorescent apoplastic tracer. Aust J Plant Physiol 15:541–555

Canny MJ (1990) What becomes of the transpiration stream? New Phytol 114:341–368

Carpita NC (1982) Limiting diameters of pores and the surface structure of plant cell walls. Science 218:813–814

Carpita NC, Gibeaut DM (1993) Structural models of primary cell walls in flowering plants: consistency of molecular structure with the physical properties of the walls during growth. Plant J 3:1–30

Carpita NC, Sabularse D, Montezinos D, Delmer DP (1979) Determination of the pore size of cell walls of living plant cells. Science 205:1144–1147

Chang PLY, Trevithick JR (1974) How important is secretion of exoenzymes through apical cell walls of fungi? Arch Microbiol 101:281–293

Clarke AE, Abbot A, Mandel TE, Pettitt JM (1980) Organization of the wall layers of the stigmatic papillae of *Gladiolus gandavensis*: a freeze-fracture study. J Ultrastruct Res 73:269–281

Clarkson DT, Robards AW (1975) The endodermis, its structural development and physiological role. In: Torrey JJ, Clarkson DT (eds) The development and function of roots. Academic Press, London, pp 415–436
Cole L, Coleman J, Evans D, Hawes C (1990) Internalisation of fluorescein isothiocyanate and fluorescein isothiocyanate-dextran by suspension-cultured plant cells. J Cell Sci 96:721–730
Currier H (1957) Callose substances in plant cells. Am J Bot 44:478–488
Ehwald R, Heese P, Klein U (1991) Determination of size limits of membrane separation in vesicle chromatography by fractionation of polydisperse dextran. J Chromatogr 542:239–245
Ehwald R, Woehlecke H, Titel C (1992) Cell wall microcapsules with different porosity from suspension cultured *Chenopodium album*. Phytochemistry 31:3033–3038
Evert RF, Derr W (1964) Callose substance in sieve elements. Am J Bot 51:552–559
Fahn A (1988) Secretory tissues in plants. New Phytol 108:229–257
Fincher GB (1989) Molecular and cellular biology associated with endosperm mobilization in germinating cereal grains. Annu Rev Plant Physiol Plant Mol Biol 40:305–346
Fry SC (1988) The growing plant cell wall: chemical and metabolic analysis. Longman, Harlow
Gogarten JP (1988) Physical properties of the cell wall of photoautotrophic suspension cells from *Chenopodium rubrum* L. Planta 174:333–339
Gubler F, Ashford AE, Jacobsen JV (1987) The release of α-amylase through gibberellin-treated barley aleurone cell walls. Planta 172:155–161
Handa AK, Bressan RA, Handa S, Hasegawa PM (1982) Characteristics of cultured tomato cells after prolonged exposure to medium containing polyethylene glycol. Plant Physiol 69:514–521
Harris N, Chaffey NJ (1986) Plasmatubules – real modifications of the plasmalemma. Nord J Bot 6:599–607
Heslop-Harrison J, MacKenzie A (1967) Autoradiography of soluble [2-^{14}C]thymidine derivatives during meiosis and microsporogenesis in *Lilium* anthers. J Cell Sci 2:387–400
Hoggart RM, Clarke AE (1984) Porosity of *Gladiolus* stigmatic papillae and pollen tube walls. Ann Bot 53:271–277
Horn MA, Heinstein PF, Low PS (1992) Characterization of parameters influencing receptor-mediated endocytosis in cultured soybean cells. Plant Physiol 98:673–679
Hoson T, Nevins DJ (1989) β-D-Glucan antibodies inhibit auxin-induced cell elongation and changes in the cell wall of *Zea* coleoptile segments. Plant Physiol 90:1353–1358
Inouhe M, Nevins DJ (1991) Inhibition of auxin-induced cell elongation of maize coleoptiles by antibodies specific for cell wall glucanases. Plant Physiol 96:426–431
Iraki NM, Bressan RA, Carpita NC (1989) Extracellular polysaccharides and proteins of tobacco cell cultures and changes in composition associated with growth-limiting adaptation to water and saline stress. Plant Physiol 91:54–61
Knox RB, Heslop-Harrison JH (1970) Direct demonstration of the low permeability of the angiosperm meiotic tetrad using a fluorogenic ester. Z Pflanzenphysiol 62:451–459
Lamport DTA, Catt JW (1981) Glycoproteins and enzymes of the cell wall. In: Tanner W, Loewus FA (eds) Encyclopedia of plant physiology, vol 13B. Plant carbohydrates. II. Extracellular carbohydrates. Springer, Berlin, Heidelberg New York, pp 133–165
Mascarenhas JP (1975) The biochemistry of angiosperm pollen development. Bot Rev 41:259–314
McCann MC, Wells B, Roberts K (1990) Direct visualization of cross-links in the primary plant cell wall. J Cell Sci 96:323–334
Meiners S, Gharyal PK, Schindler M (1991) Permeabilization of the plasmalemma and wall of soybean root cells to macromolecules. Planta 184:443–447
Meredith SC, Nathans GR (1982) Gel-permeation chromatography of asymmetric proteins. Anal Biochem 121:234–243
Mexal J, Fisher JT, Osteryoung J, Reid CPP (1975) Oxygen availability in polyethylene glycol solutions and its implications in plant-water relations. Plant Physiol 55:20–24
Mexal J, Reid CPP (1972) The growth of selected mycorrhizal fungi in response to induced water stress. Can J Bot 51:1579–1588
Michel BE, Kaufmann MR (1973) The osmotic potential of polyethylene glycol 6000. Plant Physiol 51:914–916
Money NP (1990) Measurement of pore size in the hyphal cell wall of *Achlya bisexualis*. Exp Mycology 14:234–242

O'Brien TP, Carr DJ (1970) A suberized layer in the cell walls of the bundle sheath of grasses. Aust J Biol Sci 23:275–287

O'Driscoll D, Read SM, Steer MW (1993) Determination of cell-wall porosity by microscopy: walls of cultured cells and pollen tubes. Acta Bot Neerl 42:237–244

Owen TP Jr, Thomson WW (1991) Structure and function of a specialized cell wall in the trichomes of the carnivorous bromeliad *Brocchinia reducta*. Can J Bot 69:1700–1706

Paull RE, Jones RL (1976) Studies on the secretion of maize root cap slime. V. The cell wall as a barrier to secretion. Z Pflanzenphysiol 79:154–164

Paull RE, Jones RL (1978) Regulation of synthesis and secretion of fucose-containing polysaccharides in cultured sycamore cells. Aust J Plant Physiol 5:457–467

Peters R (1986) Fluorescence microphotolysis to measure nucleocytoplasmic transport and intracellular mobility. Biochim Biophys Acta 864:305–359

Peterson CA, Emanuel ME, Humphreys GB (1981) Pathway of movement of apoplastic fluorescent dye tracers through the endodermis at the site of secondary root formation in corn (*Zea mays*) and broad bean (*Vicia faba*). Can J Bot 59:618–625

Preston RA, Murphy RF, Jones EW (1987) Apparent endocytosis of fluorescein isothiocyanate-conjugated dextran by *Saccharomyces cerevisiae* reflects uptake of low molecular weight impurities and not dextran. J Cell Biol 105:1981–1987

Richter E, Ehwald R (1983) Apoplastic mobility of sucrose in storage parenchyma of sugar beet. Physiol Plant 58:263–268

Robards AW, Lucas WJ (1990) Plasmodesmata. Annu Rev Plant Physiol Plant Mol Biol 41:369–419

Robinson DG, Hillmer S (1990) Endocytosis in plants. Physiol Plant 79:96–104

Rodríguez-García MI, Majewska-Sawka A (1992) Is the special callose wall of microsporocytes an impermeable barrier? J Exp Bot 43:1659–1663

Rougier M (1981) Secretory activity of the root cap. In: Tanner W, Loewus FA (eds) Encyclopedia of plant physiology, vol 13B. Plant carbohydrates. II. Extracellular carbohydrates. Springer, Berlin, Heidelberg New York, pp 542–574

Rowley JR, Dunbar A (1970) Transfer of colloidal iron from sporophyte to gametophyte. Pollen Spores 12:305–328

Sauter JJ, Marquardt H (1970) Cytochemical investigations on cytochrome oxidase and succinic dehydrogenase activity in pollen tetrads. Z Pflanzenphysiol 63:15–18

Shedletzky E, Shmuel M, Trainin T, Kalman S, Delmer D (1992) Cell wall structure in cells adapted to growth on the cellulose-synthesis inhibitor 2,6-dichlorobenzonitrile. Plant Physiol 100:120–130

Squire PG (1981) Calculation of hydrodynamic parameters of random coil polymers from size exclusion chromatography and comparison with parameters by conventional methods. J Chromatogr 210:433–442

Stevenson TT, McNeil M, Darvill AG, Albersheim P (1986) Structure of plant cell walls. XVIII. An analysis of the extracellular polysaccharides of suspension-cultured sycamore cells. Plant Physiol 80:1012–1019

Taylor IEP, Wallace JC, MacKay AL, Volke F (1990) Use of chemical fractionation and proton nuclear magnetic resonance to probe the physical structure of the primary plant cell wall. Plant Physiol 94:174–178

Tepfer M, Taylor IEP (1981) The permeability of plant cell walls as measured by gel filtration chromatography. Science 312:761–763

Wallace JC, MacKay AL, Sasaki K, Taylor IEP (1993) A proton nuclear magnetic resonance study of the physical changes in growing plant cell walls. Planta 190:227–230

Analysis of Chitin Biosynthesis

MICHEL FÈVRE, LUCIEN GAY, and HENRI CHANZY

1 Introduction

The cell wall in fungi fulfills a number of vital functions. One of its obvious roles is to protect the protoplast by constituting a structural barrier. Often considered as a static structure, the wall is submitted to modification and remodeling during the fungal life cycle as attested by qualitative and quantitative analysis. Because of its rigidity, the wall maintains the shape of the cell and it is the association of microfibrillar polysaccharides (chitin and/or cellulose) embedded in an amorphous matrix of β-glucans which confers that rigidity.

The most characteristic component, chitin, has now been found in all classes of higher fungi (Bartnicki-Garcia 1968), including fungi traditionally regarded as cellulosic fungi (Bulone et al. 1992) and chitin microfibrils suspected to form the main skeletal element of the walls. Due to its predominant role, and as it is not formed by plant or mammalian tissues, chitin has been considered as a target for antifungal agents. Compounds such as polyoxins and nikkomycins, which inhibit chitin biosynthesis, are efficient antifungal agents (Endo et al. 1970). Defense-related phenomena induced by pathogen attack of higher plants include the induction of pathogenesis-related proteins (PR). Several PR proteins have been identified as β-1,3-glucanases and chitinases which inhibit in vitro fungal growth (Sahai and Manocha 1993).

The understanding of the synthesis of chitin at the biochemical and molecular level could allow the determination of the mechanisms which control chitin synthetase.

2 Structure of Chitin

Chitin is a homopolymer of N-acetyl glucosamine units linked by β-1,4 glycosidic bonds. The polymer is a linear chain in which the sugar units are in the chair form and each monomer is inverted 180° with respect to its neighbour; thus, a disaccharide is the repeating unit in the chain. The chains associate with one another by hydrogen bondings, 0.0 between the 0 group of C3 of one unit and the 0 ring of its neighbour in the chain, CO. NH in adjacent chains. This structure makes a strong rigid polymer and accounts for the formation of microfibrils.

Modern Methods of Plant Analysis, Vol. 17
Plant Cell Wall Analysis
Edited by H.-F. Linskens and J.F. Jackson

Chitin chains are often associated with microfibrils which can be easily revealed by electron microscopy.

Several crystalline forms have been identified: α-chitin in which adjacent chains are arranged in antiparallel forms, i.e., facing alternatively in opposite directions. This contrasts with β-chitin in which the chains are parallel. In fungi, chitin is present in the α-form, while the β-form is detected in other organisms (Blackwell 1969; Minke and Blackwell 1978).

3 The Enzymatic Synthesis of Chitin

The first evidence for chitin biosynthesis in vitro was obtained by Glaser and Brown (1957) who demonstrated that uridine diphosphate N-acetyl glucosamine (UDP-GlcNAc) was the precursor of the synthesis of a polymer recognized as chitin by its unsolubility and its digestibility with chitinase. The ability of UDP-GlcNAc to serve as a precursor of chitin synthesis in cell-free extracts from fungal species from all taxonomic classes (Table 1) has been determined (for review, see Cabib 1987).

Table 1. Properties of the chitin synthases from representative fungi of the main taxonomic classes

	pH	Cation stimulation	UDPGlc NAc K_m (mM)	Protease stimulation	Polyoxin inhibition k_i (μM)	References
Cellulosic fungi						
Oomycetes *Saprolegnia monoica*	7.5	Mg^{2+}	0.2	+	20	Gay et al. (1993a)
Chitinous fungi						
Zygomycetes *Mucor rouxii*	6.5	Mg^{2+}	0.5	+	6.5	Bartnicki-Garcia and Lippman (1972)
Ascomycetes *Neurospora crassa*	8	Mg^{2+}	2.1	+	1,4	Endo et al. (1970)
Hemiascomycetes *Saccharomyces cerevisiae*	6.5	Mg^{2+}	0.5	+	1	Bowers et al. (1974)
Basidiomycetes *Coprinus cinereus*	7.5	Mg^{2+}	1	–	3	Gooday and Rousset-Hall (1975)

3.1 Assay of Chitin Synthase

The assay is usually based on the incorporation of carbon 14-labeled (in the amino sugar) substrate into the insoluble, radioactive product. Mg^{2+} and N-acetyl glucosamine are also present in the reaction mixture. After stopping the reaction, using 10% trichloracetic acid, N-acetic acid, or 95% ethanol, the mixture is filtered on glass fiber filters and radioactivity of the remaining insoluble product on the filter is determined by scintillation spectrometry.

The stoichiometry of the reaction as determined with the enzyme from *S. cerevisiae* is (Cabib 1987):

$$2\text{n UDP} - \text{GlcNAc} \rightarrow 2\text{n UDP} + \left(\text{GlcNAc}\ \beta - \left(1-4\right)\ \text{GlcNa}\right)\text{n}.$$

A general activator of the activity is free N-acetyl glucosamine (GlcNAc), which can stimulate the substrate incorporation up to tenfold depending on the fungal enzyme preparation (Cabib 1987). This amino sugar is not incorporated into chitin and it is unlikely that it functions as a primer, but rather as an allosteric activator (Horsh and Rast 1993). Another general universal characteristic is the requirement for a divalent cation; Mg^{2+} (at 1.5 mM) is commonly the most effective, but Mn^{2+} and Co^{2+} are also active in some cases (Choi and Cabib 1994).

The enhancement of activity using partial proteolysis is also a general property of fungal chitin synthase (Cabib et al. 1979). Trypsin pretreatment of the chitin synthases or addition of protease to the assay mixtures stimulates substrate incorporation up to 30-fold. Activation of chitin synthase by protease is not simply due to unmasking membrane-bound enzymes, as detergent-solubilized enzymes are still activated by trypsin (Cabib et al. 1979; Gay et al. 1993a). The enzyme seems to be synthesized as an inactive form, i.e., zymogen, which required a moderate proteolytic attack to be activated (Cabib et al. 1979).

Assay conditions have been established to measure the total chitin synthase activity from cell-free extracts. However, molecular genetics has revealed that fungi contain several genes encoding different chitin synthases (Bulawa 1993). Gene disruptions have allowed the creation of defective mutants and revealed the enzymatic activity of chitin synthases which were previously undetected in crude enzyme preparations because of their relatively low activity (Sburlati and Cabib 1986; Orlean 1987). Using these strains with chitin synthase disruptions, conditions (cation requirement, cation inhibition, optimum pH) for measurement of each chitin synthase activity have been set up in yeast (Table 2) which allow the establishment of assay conditions for the specific determination of each enzyme in the presence of the others (Choi and Cabib 1994). These assay conditions can thus serve as a basis to test chitin synthase activities in other fungi.

3.2 Inhibition of Enzymatic Activity

Chitin synthases are competitively inhibited by polyoxins and nikkomycin (Endo et al. 1970; Gooday 1989). These structural analogs of the substrate UPD-GlcNAc

Table 2. Assay conditions for chitin synthases from *Saccharomyces cerevisiae*. Assays conditions have been established for the determination of each synthetase in the presence of the others. (Choi and Cabib 1994)

Chitin synthase	pH	Divalent cation	Protease activation
CHS 1	6.5	Mg^{2+}	+
CHS 2	8	Co^{2+}	+
CHS 3	7.5–8	Co^{2+}, Ni^{2+}	–

behave as classical competitive inhibitors with constant inhibition values of 10^{-6}–10^{-7}. These inhibitors can be used to ascertain in vitro chitin synthase activity even if in vivo the organisms are refractory to their actions.

4 Preparation of Chitin Synthase

Chitin synthases are membrane-bound enzymes. Using differential centrifugation and density gradient centrifugation of cell-free extracts, the enzymes have been found to be attached to the plasma membrane (Duran et al. 1979). The enzyme appears to be an intrinsic protein leading to a vectorial synthesis of chitin by accepting substrate, cations and effectors from the cytoplasm and extruding chitin chains to the cell wall.

However, specific chitin synthesizing particles have been isolated by a sucrose density gradient, and electron microscopy has revealed the vesicular structure of these particles, which have been called chitosomes (Bartnicki-Garcia et al. 1979). Using a sophisticated procedure for isolation, chitosomes have been identified in all the major groups of chitinous fungi (Bartnicki-Garcia et al. 1978). Chitosomes may represent the microvesicular vehicle delivering the enzyme to the plasma membrane. However, cellulosic fungi such as *Saprolegnia*, which have been shown to synthesize chitin (Gay et al. 1993b), do not seem to contain chitosomes (Gay and Bartnicki-Garcia, unpub. results). This indicates that chitin synthase may be secreted via a different type of vesicle in cellulosic fungi. With improved methods for separating membranous organelles by density gradient centrifugation, chitin synthase has been demonstrated to be present in both chitosomes and plasma membranes. Moreover, different enzymes have the same subcellular localization, e.g. in *S. cerevisiae*, chitin synthases 1 and 2 are associated with the same membranes (Leal-Morales et al. 1994). Plasma membrane-bound and chitosome-entrapped chitin synthases are mainly in a zymogenic form and can be activated by limited proteolysis.

Chitin synthases have been solubilized from membranes using digitonin (Cabib et al. 1979; Rousset-Hall and Gooday 1975), but CHAPS and octylglucoside are also efficient detergents to solubilize the enzymes (Gay et al. 1993a). As chitin synthases are membrane-bound, their activity is influenced by a phospholipidic environment and some solubilized enzymes require phospholipids (Machida et al. 1994). Digitonin is also able to dissociate chitosome into small particles with a sedimentation coefficient of 16 s (Bartnicki-Garcia et al. 1979). This is equivalent to that of spherical proteins of MW 500 000. Interestingly, solubilized enzymes partially purified from plasma membranes were found to have a MW of about 570 000 similar to the chitosome subunits (Cabib 1987).

In spite of numerous attempts to achieve purification of chitin synthases, these enzymes have not been convincingly obtained in a pure form. Machida and Saito (1993) reported the purification of a chitin synthase from the filamentous fungus *Absidia glauca*. The purified enzyme existed as a zymogen and gave a single band with a MW of 30 000. Surprisingly, the amino-terminal sequence of the purified polypeptide had complete homology with the lectin Con A. It can be noted that the molecular weights of the proteins encoded by the *chs* genes cloned from yeasts and filamentous fungi are over 100 000 (Bulawa 1993). As the native chitin synthases (solubilized enzymes or chitosome-associated enzymes) have a MW of about 500 000, it is most probable that five or six identical proteins combine to form complexes capable of synthesizing the chitin chains, which assemble into microfibrils. One can speculate that these enzymatic complexes will have a structural organization similar to the "rosette" from higher plants involved in cellulose biosynthesis.

5 Product Characterization of Chitin Synthase

5.1 Chemical Identification

Insoluble (14C) reaction products collected by centrifugation (48 000 × *g*, 30 min) can be shown to be chitin by means of HCl hydrolysis and chitinase digestion. The products can be hydrolyzed under vacuum in 6 M HCl at 120 °C for 2.5 h. (They should be insoluble after treatments at 100 °C for 1 h in 1 KOH or 1 M acetic acid.) After removal of HCl by a stream of N_2, the hydrolysate is spotted on Whatman paper and submitted to descending chromatography using propanol, ethyl acetate/water (7 : 1 : 2 v/v) as the solvent. The spot of hydrolysate corresponding to the position of the authentic acetyl glucosamine can be excised and the radioactivity measured.

For enzyme hydrolysis, purified chitinase dissolved in 2 mM sodium phosphate buffer, pH 6, is added to the insoluble reaction product (antibiotic can be added to prevent development of bacteria during prolonged incubation). After up to 20 h of incubation, the mixture is subjected to TLC chromatography on cellulose plates using butanol/pyridine/water (6 : 4 : 3 v/v) as a solvent. The chitinase releases diacetylchitobiose and GlcNAc, which can be revealed by spraying with 0.3% p-aminohippuric acid in phosphoric acid/80% ethanol (0.5 : 100 v/v). Bands corresponding to the position of standard sugars can be scraped from the chromatograms to measure radioactivity.

5.2 Biophysical Identification of Chitin

Electron microscopy studies have revealed that the insoluble polymer synthesized in vitro is in the form of spindle-like particles or microfibrils (Bartnicki-Garcia et al. 1979; Duran et al. 1979; Gay et al. 1992). The reaction product can be collected by centrifugation extracted with 5% (w/v) aqueous KOH solution, washed, then used for biophysical characterization and compared to α-chitin extracted from arthropod cuticles or to chitin extracted from hyphal cell walls. By combining electron microscopy and biophysical analysis, it is possible to obtain the real morphology and organization of the products synthesized in vitro taking advantage of their being a crystalline polymer.

5.2.1 Fourier Transform Infrared Spectroscopy

For this method of analysis, thin films of reaction products can be prepared by allowing the aqueous suspension of the reaction products to evaporate. The film obtained can be mounted on a specimen holder and Fourier transform infrared (FTIR) spectra can be obtained in transmission mode. For instance, we used a Perkin Elmer 1720 × FTIR instrument equipped with a microfocus accessory (Gaill et al, 1992). This method allowed the characterization of the products from enzymes of various fungi (Gooday 1979; Bulone et al. 1992).

Figure 1 shows that the infrared spectra of chitin, either synthesized in vitro by enzymes from *Saprolegnia monoica* or in vivo from the cell wall of *S. monoica* protoplasts, are nearly identical to those of a chitin sample chemically purified from crab tendon. In particular, they display three C=O absorption bands near 1560, 1622, and 1656 cm^{-1} that are characteristic of α-chitin (Pearson et al. 1960). On the other hand, these spectra differ from those of β-chitin, which display only two absorption bands near 1560 and 1626 cm^{-1} (Falk et al. 1966).

5.2.2 X-Ray Diffraction Analysis

As the polymers synthesized in vitro have a morphology, i.e., microfibrils having spindle-like particles, similar to that of the crystalline polymer extracted from cell walls, this suggests that these polymers may have similar crystallographic structures. This similarity can be revealed by diffraction analysis.

For instance, X-ray diffraction analysis can be used to identify various chitins from in vitro products by comparison with well-characterized chitin standards. In our setup, the various samples are inserted in thin X-ray glass capillaries that are

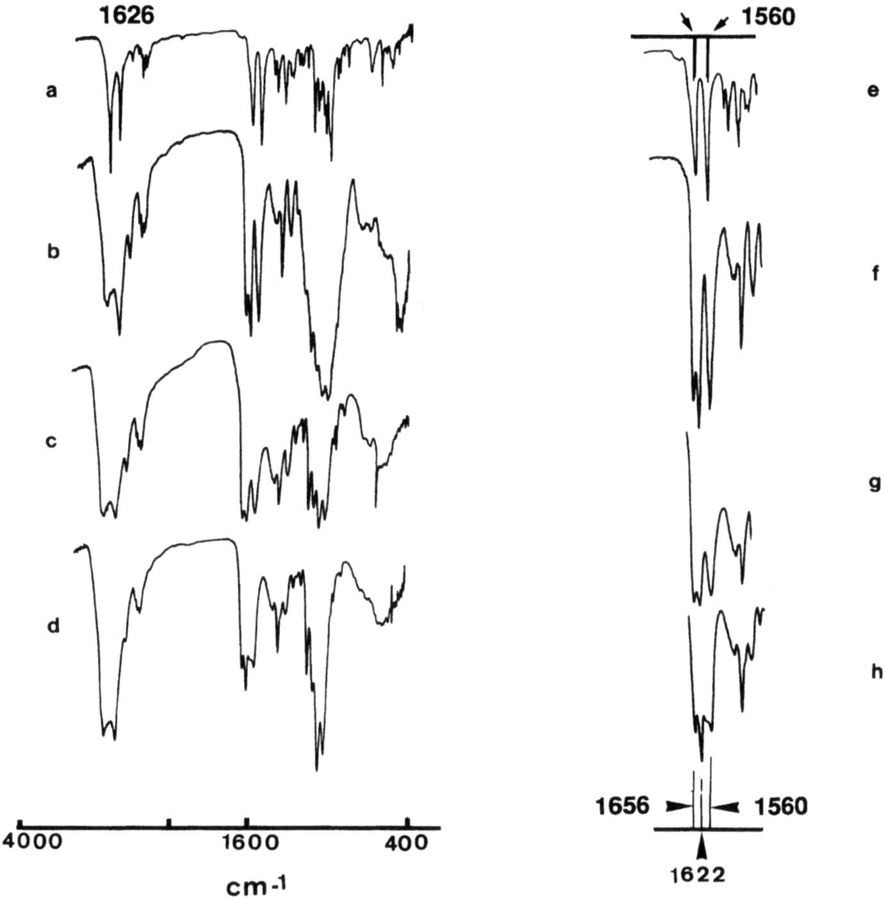

Fig. 1a–h. Comparison of the FTIR spectra of chitin originating from *S. monoica* with those of α and β chitin standards. **a–d** Low resolution spectra chitin from a vestimentiferan tubes (Gaill et al. 1992); **b** a sample of single crystals of α chitin (Persson et al. 1992); **c** a sample of in vitro chitin synthesized by solubilized enzyme from *S. monoica* (Gay et al. 1993b); **d** chitin from the wall of protoplasts of *S. monoica* (Bulone et al. 1992). **e–g**, as in **a–d**, respectively, but showing the enlargement of the fraction of the carbonyl part of the spectra

mounted in a flat-film Warhus vacuum camera and subjected to Cuka radiation. Typically, the in vitro chitin, synthesized by the enzymes from *S. monoica*, presents reflections at 0.943, 0.516, 0.466 and 0.341 nm (Fig. 2C). These spacings correspond exactly to those of an α-chitin sample extracted from lobster tendon (Fig. 2B). On the other hand, they are quite different from those of β-chitin from diatom spines (Fig. 2A). A sample of chitin from the wall of protoplasts from *S. monoica* (Fig. 2D) also presented the same diffraction features as both in vitro or lobster tendon chitin.

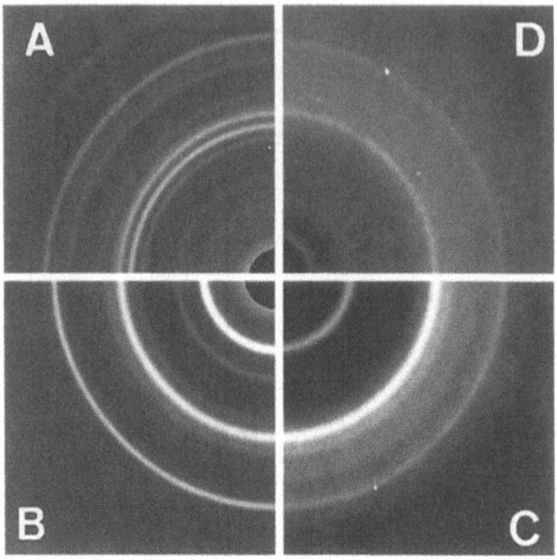

Fig. 2A–C. Typical X-ray diagrams of **A** a sample of β chitin from the spines of *Thalassiosira fluviatilis*; **B** a sample of α chitin from lobster tendon; **C** a chitin sample resulting from the in vitro biosynthesis of chitin using the chitin synthase from *S. monoica*. **D** A chitin sample corresponding to purified wall of *S. monoica* protoplasts

5.2.3 Electron Microscopy and Electron Diffraction Analysis

By combining the imaging and diffraction mode of the transmission electron microscope (TEM), it is possible to characterize easily the various chitin samples prepared either in vitro or in vivo. This characterization can be typically achieved with a transmission electron microscope Philips EM 400 T operated at 120 kV under a reduced electron dose. Figure 3A shows the fibrillar network of the wall of a disencrusted protoplast of *S. monoica* (Bulone et al. 1992). This network can be characterized by an electron diffraction technique. The insert in Fig. 3A presents a typical electron diffractogram corresponding to the circled area. It displays two sharp diffraction rings: one with a strong intensity at 0.465 nm and another of weaker intensity at 0.339 nm. These two rings, which disappear under strong electron irradiation, are characteristic of α-chitin. In Fig. 3B, a typical specimen of chitin extracted from the wall of *S. monoica* is shown. This sample consists of a very fine granular structure that diffracts also as α-chitin, identified by the two strong diffraction rings (Fig. 3B, insert) at 0.339 and 0.465 nm.

Another form of biosynthetic chitin is presented in Fig. 3C. This electron micrograph correponds to a sample resulting from the in vitro synthesis of chitin using solubilized chitin synthases isolated from *S. monoica* (Gay et al., 1993a). Here, the sample is observed after negative staining with uranyl acetate. It consists of small spindle-like elements that can also be characterized as α-chitin by biophysical methods, such as X-ray and electron diffraction analysis as well as Fourier transform infrared spectroscopy. In this case, each element is, in fact, a small crystal of α-chitin and it can be shown by the microdiffraction method that within

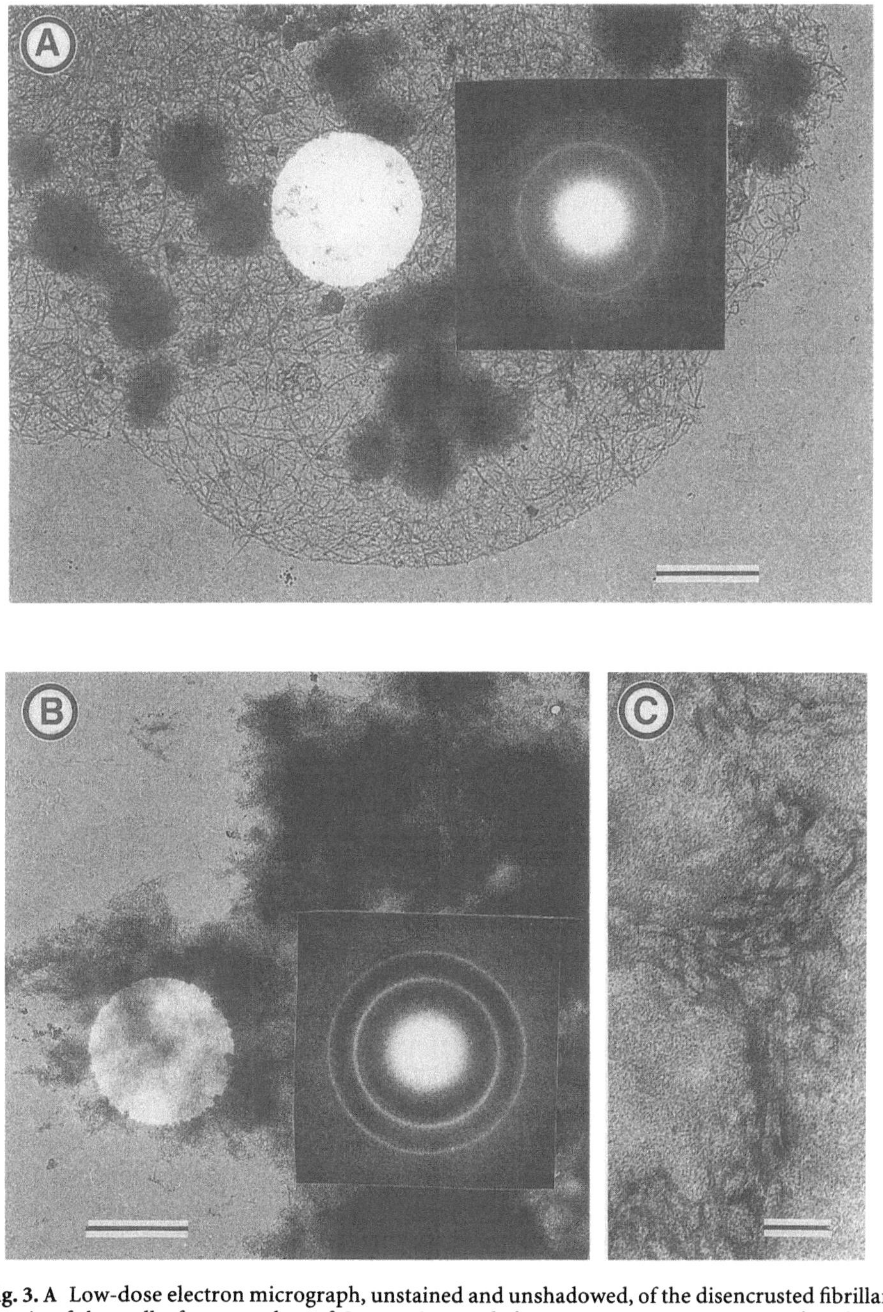

Fig. 3. A Low-dose electron micrograph, unstained and unshadowed, of the disencrusted fibrillar matrix of the wall of a protoplast of *S. monoica*. *Scale bar* =1 μm. *Insert* in **A**, Typical α chitin electron diffraction diagram corresponding to the *circled area*. **B**, as in **A**, but corresponding to the chitin component of the wall of *S. monoica*. *Scale bar* =1 μm. *Insert* in **B**, Typical electron diffraction diagram of the *circled area*. **C** In vitro chitin synthesized by solubilized chitin synthase extracted from *S. monoica*. The specimen was negatively stained with uranyl acetate. *Scale bar* =50 nm

each crystal the chitin chain axis is oriented along the length of the spindle (Bartnicki-Garcia et al. 1994).

6 Polymerization and Crystallization

In vivo, dyes such as Calcofluor or Congo red were shown to bind to glycan chains, preventing the assembly of chitin or cellulose microfibrils produced by bacteria and algae (Haigler and Benziman 1982; Herth and Schnepf 1982). It was concluded that the processes of chain polymerization and chain crystallization do not occur simultaneously and that the dyes uncoupled these two processes.

Electron microscopy and electron diffraction have been used to elucidate the effects in vitro of the dyes on chitin biogenesis by purified chitosomes (Bartnicki Garcia et al. 1994). Chitin synthase activity was inhibited by the dyes and both polymerization and chain assembly were affected. This indicates that in vitro the dyes have access to the synthesizing enzymes, inhibiting enzymatic activity, while in vivo they affect only the extrusion of the polymer chains. Electron diffraction studies revealed that both dyes, which greatly interfered with polymer crystallization, induced the production of gel-like precipitates instead of the usual microfibrillar products. Stacks of these dyed molecules were adsorbed on the chitin macromolecules, preventing the lateral growth of chitin crystals (Bartnicki-Garcia et al. 1994).

7 Fungal Chitin Synthase Genes

Genes encoding chitin synthase activity have been cloned from several yeasts and filamentous fungi (for a review, see Bulawa 1993). Alignment of the deduced amino acid sequences of three chitin synthase genes, *Saccharomyces cerevisiae chs1* and *chs2* and *Candida albicans chs1*, reveals that the enzymes contained highly conserved sequences (Bowen et al. 1992). Degenerate polymerase chain reaction primers, encoding short, completely conserved regions, were developed. These primers were first used to amplifiy 600-bp fragments homologous to chitin synthase from the genomic DNA of 14 fungal species (Bowen et al. 1992). Their use has now been extended to numerous other fungal species (Yarden and Yanofsky 1991; Mehman et al. 1994). Comparison of the deduced amino acid sequences of the amplified DNA fragments revealed that the sequences fell into three classes which could represent different functional groups (Bowen et al. 1992). The *chs* sequences support the phylogenetic analysis of ectomycorrhizal fungi, which confirmed the current taxonomic groups; moreover, these sequences may have potential use in fungal identification (Mehman et al. 1994).

Catalysis by most glycosyl transferases must take place by an “inversion” mechanism, since the donor substrates are α-linked nucleotide diphospho-sugars

(UDP-N-acetylglucosamine, UDP-glucose) from which the sugar is transferred to the growing molecule forming a β-linked product (chitin, cellulose).

Recently, hydrophobic cluster analysis (HCA) and comparison of the amino acid sequences of a number of UDP glycosyl transferases and polysaccharide synthases, including chitin synthase, have revealed that these enzymes share a modular architecture with transmembrane segments attached to a globular region composed of two conserved domains. Domain A is common to all glycosyl transferases and glycosyl transferases, while the polymerases have, in addition, domain B (Saxena et al. 1995). Glycosyl transferases, which transfer one sugar, contain only domain A. The proteins having two catalytic domains (A, B) are responsible for the dual addition of glycosyl residues necessary for the processive reaction and synthesis of the polymer. The simultaneous or sequential addition of two monosaccharides allows the twofold screw axis to be generated without rotation of the polymer chain or the protein. This is of particular importance, e.g. in the case of chitin or cellulose, where a number of polymer chains are simultaneously synthesized and assembled in crystalline microfibrils. This double addition, which involved opposite (facing) catalytic domains in the synthase, seems to be ideal for the synthesis of chitin and also cellulose where a disaccharide is the repeating unit of the chain (Saxena et al. 1995).

8 Morphogenetic Roles of Chitin Synthases

As chitin is an important structural component of many fungal cell walls, control of chitin synthesis and degradation must be an important aspect of growth. Three chitin synthase genes of the yeast *Saccharomyces cerevisiae* have been isolated and characterized. *Chs1*- and *chs2*-encoded proteases activated chitin synthase zymogens and the enzymes encoded by *CAL1/chs3* did not require protease activation (Cabib 1987; Bulawa 1993; Choi and Cabib 1994). Molecular genetics has provided evidence for the role played by the enzymes during growth. In gene disruption experiments, each of the genes was shown to be nonessential for cell growth. However, abnormal cell morphologies were induced. From these abnormalities, caused by the mutation of each of these genes, it has been demonstrated that CHS3 is responsible for the formation of most of the cellular chitin. CHS2 is important for primary septum formation, while CHS1 acts as a repair enzyme (Bulawa et al, 1986; Cabib et al. 1989; Shaw et al. 1991). Gene expression studies have confirmed the supposed role of the genes, as the levels of transcription of each of the chitin synthase genes increased at that stage of the cell cycle in which they are supposed to function (Choi and Cabib 1994; Pammer et al. 1992).

In filamentous fungi, there is more than one chitin synthase gene (Yarden and Yanofsky 1991; Bowen et al. 1992; Chua et al. 1994; Gold and Kronstad 1994). However, there is little information on the function of the genes. Inactivation of two chitin synthase genes from *Neurospora crassa* and *Ustilago maydis* induced reduced chitin synthase activity (as measured in vitro), but the disrupted strains were still viable, suggesting the presence of other genes (Din and Yarden 1994; Gold and Kronstad 1994).

9 Conclusions

Genetic techniques have increased our understanding of chitin biosynthesis and allow speculation about the mechanism of chitin synthesis and the roles played by the protein encoded by the different genes. However, our knowledge of the protein encoded by the genes is still limited. One unsolved problem is the fact that these membrane-bound proteins have not yet been purified, however, it is expected that gene fusion technology and purification of recombinant proteins will allow the study of the organization of the chitin synthase complexes. Most fungal species seem to contain several *chs* genes which assume different functions during growth. Electron microscopy and biophysical analysis of the products synthesized by these different chitin synthases have not yet been performed. Using gene disruption to create specific defective mutants, it will be of interest to determine whether the morphology and structure of the chitin synthesized by each enzyme are related to its supposed specific morphogenetic function.

References

Bartnicki Garcia S (1968) Cell wall chemistry, morphogenesis and taxonomy Annu Rev Microbiol 22:87–108

Bartnicki-Garcia S, Lippman E (1972) Inhibition of *Mucor rouxii* by polyoxin effects on chitin synthetase and morphological development. J Gen Microbiol 71:301–309

Bartnicki-Garcia S, Bracker CE, Reyes E, Ruiz-Herrera J (1978) Isolation of chitosomes from taxonomically diverse fungi and synthesis of chitin microfibrils in vitro. Exp Mycol 2:173–192

Bartnicki-Garcia S, Ruiz-Herrera J, Bracker CE (1979) Chitosomes and chitin synthesis. In: Burnett JH, Trinci APJ (eds) Fungal walls and hyphal growth. Cambridge University Press, Cambridge, pp 149–168

Bartnicki Garcia S, Persson J, Chanzy H (1994) An electron microscope and electron diffraction study of the effect of calcofluor and Congo red on the biosynthesis of chitin. Arch Biochem Biophys 310:6–15

Blackwell J (1969) Structure of β chitin or paralled chain systems of poly β-(1–4)N-acetylglucosamine. Biopolymers 7:281–298

Bowen AR, Chen-Wu JL, Momany M, Young R, Szaniszlo PJ, Robbins PN (1992) Classification of fungal chitin synthases. Proc Natl Acad Sci USA 89:519–523

Bowers B, Levib G, Cabib E (1974) Effect of polyoxin D on chitin synthase and septum formation in *Saccharomyces cerevisiae*. J Bacteriol 119:564–575

Bulawa CE (1993) Genetics and molecular biology of chitin synthesis in fungi. Annu Rev Microbiol 47:505–534

Bulawa CE, Slater M, Cabib E, Au-Young J, Sburlati A, Adair WL, Robbins PW (1986) The *S. cerevisiae* structural gene for chitin synthase is not required for chitin synthesis in vivo. Cell 46:213–225

Bulone V, Chanzy H, Gay L, Girard V, Fèvre M (1992) Characterization of chitin and chitin synthase from the cellulosic cel wall fungus *Saprolegnia monoica.* Exp Mycol 16:8–21
Cabib E (1987) The synthesis and degradation of chitin. Adv Enzymol 59:59–101
Cabib E, Duran A, Bowers B (1979) Localized activation of chitin synthetase in the initiation of yeast septum formation. In: Burnett JH, Trinci APJ (eds) Fungal walls and hyphal growth. Cambridge University Press, Cambridge, pp 190–201
Cabib E, Silverman SJ, Sburlati A, Slater ML (1989) Chitin synthesis in yeast (*Saccharomyces cervisiae*). In: Kuhn PJ, Trinci APJ, Jung MJ, Goosey MW, Copping LG (eds) Biochemistry of cell walls and membranes in fungi. Springer, Berlin Heidelberg New York, pp 31–41
Choi WJ, Cabib E (1994) The use of divalent cations and pH for the determination of specific yeast chitin synthetases. Anal Biochem 219:368–372
Choi WS, Santos M, Duran A, Cabib E (1994) Are yeast chitin synthases regulated at the transcriptional or the post transcriptional level? Mol Cell Biol 14:7685–7694
Chua SS, Momamy M, Mendoza L, Szaniszlo PJ (1994) Identification of three chitin synthase genes in the amorphic fungal pathogen *Sporothrix shenckii.* Curr Microbiol 29: 151–156
Din AB, Yarden O (1994) The *Neurospora crassa chs-2* gene encodes a non essential chitin synthase. Microbiology 140:2189–2197
Duran A, Cabib E, Bowers B (1979) Chitin synthetase distribution on the yeast plasma membrane. Science 203:363–365
Endo A, Kakiki K, Misato T (1970) Mechanism of action of the antifungal agent polyoxin D. J Bacteriol 104:189–196
Falk M, Smith DG, McLachlan J, McInnes AG (1966) Studies on chitin (β-(1–4)-linked 2-acetamido-2-deoxy-D-glucan) fibers of the diatom *Thalassiosira fluviatilis* Hustedt. Can J Chem 44:2269–2281
Gaill F, Persson J, Sugiyama J, Vuong R, Chanzy H (1992) The chitin system in the tubes of deep sea hydrothermal vent worms. J Struct Biol 109:116–128
Gay L, Bulone V, Girard V, Fèvre M, Chanzy H (1992) Chitin pleomorphism in the cellulosic cell wall fungus *Saprolegnia.* FEMS Microbiol Lett 100:405–410
Gay L, Chanzy H, Bulone V, Girard V, Fèvre M (1993a) Synthesis in vitro of crystalline chitin by a solubilized enzyme from the cellulosic fungus *Saprolegnia monoïca.* J Gen Microbiol 139:2117–2122
Gay L, Chanzy H, Bulone V, Girard V, Fèvre M (1993b) Biosynthesis of chitin in the fungus *Saprolegnia.* In: Muzzarelli RAR (ed) Chitin enzymology. Eur Chitin Soc Lyon, Ancona, pp 57–66
Giraud-Guille MM, Chanzy H, Vuong R (1990) Chitin crystals in arthropod cuticles revealed by diffraction contrast transmission electron microscopy. J Struct Biol 103:232–240
Glaser I, Brown DH (1957) The synthesis of chitin in cell-free extracts of *Neurospora crassa.* J Biol Chem 228:729–742
Gold SE, Kronstad JW (1994) Disruption of two genes for chitin synthase in the phytopathogenic fungus *Ustilago maydis.* Mol Microbiol 11:897–902
Gooday GW (1979) Chitin synthesis and differentiation in *Coprinus cinereus.* In: Burnett JH, Trinci APJ (eds) Fungal walls and hyphal growth. Cambridge University Press, Cambridge, pp 209–223
Gooday GW (1989) Inhibition of chitin metabolism. In: Kuhn PJ, Trinci APJ, Jung MJ, Goosey MW, Copping LG (eds) Biochemistry of cell walls and membranes in fungi. Springer, Berlin Heidelberg New York, pp 61–79
Gooday GW, de Rousset-Hall A (1975) Properties of chitin synthase from *Coprinus cinereus.* J Gen Microbiol 89:137–145
Haigler CH, Benziman M (1982) Biogenesis of cellulose I microfibrils occurs by cell directed assembly in *Acetobacter xylinum.* In: Malcom Brown R Jr (ed) Cellulose and other natural polymers. Plenum Press, New York, pp 273–297
Herth W, Schnepf E (1982) Chitin fibril formation in algae. In: Malcom Brown R Jr (ed) Cellulose and other natural polymers. Plenum Press, New York, pp 185–206
Horsch M, Rast DM (1993) Allosteric activation of chitin synthase by N-acetylglucosamine: a mechanistic study. In: Muzzarelli RAR (ed) Chitin enzymology. Eur Chitin Soc, pp 47–56

Leal Morales CA, Bracker CE, Bartnicki-Garcia S (1994) Subcellular localization, abundance and stability of chitin synthetases 1 and 2 from *Saccharomyces cerevisiae.* Microbiology 140:2207–2216

Machida S, Saito M (1993) Purification and characterization of membrane bound chitin synthase. J Biol Chem 268:1702–1707

Machida S, Todoriki S, Hamamatsu S, Saito M (1994) Phospholipid requirements of membrane bound chitin synthase from *Absidia glauca.* FEMS Microbiol Lett 115:235–240

Mehmann B, Brunner I, Brauss GH (1994) Nucleotide sequence variation of chitin synthase genes among ectomycorrhizal fungi and its potential to use in taxonomy. Appl Environm Microbiol 60:3105–3111

Minke R, Blackwell J (1978) The structure of α chitin. J Mol Biol 120:167–182

Orlean P (1987) Two chitin synthases in *Saccharomyces cerevisiae.* J Biol Chem 262:5733–5739

Pammer M, Briza A, Schuster T, Stucka R, Feldmann H, Breitenbach M (1992) *DIT 101* (*CSD2*, *CAL1*), a cell cycle regulated yeast gene required for synthesis of chitin in cell walls and chitosan in spore walls. Yeast 8:1089–1099

Pearson FG, Marchessault RH, Liang CY (1960) Infrared spectra of crystalline polysaccharide. V. Chitin. J Polymer Sci 53:101–116

Persson JE, Domard A, Chanzy H (1992) Single crystals of α-chitin. Int J Biol Macromol 14:221–224

Rousset-Hall A, Gooday GW (1975) A kinetic study of solubilized chitin synthetase preparation from *Coprinus cinereus.* J Gen Microbiol 89:146–159

Sahai AS, Manocha MS (1993) Chitinases of fungi and plants: their inolvement in morphogenesis and host-parasite interactions. FEMS Microbiol Rev 11:317–338

Saxena IM, Brown RM Jr, Fèvre M, Geremia R, Henrissat B (1995) Multidomain architecture of glycosyl transferases: implications for mechanism of action. J Bacteriol 177:1418–1424

Sburlati A, Cabib E (1986) Chitin synthase 2, a presumptive participant in septum formation in *Saccharomyces cerevisiae.* J Biol Chem 261:15147–15152

Shaw JA, Mol PC, Bowers B, Silverman SJ, Valdivieso MH, Duran A, Cabib E (1991) The function of chitin synthase 2 and 3 in the *Saccharomyces cerevisiae* cell cycle. J Cell Biol 114:111–123

Yarden O, Yanofsky C (1991) Chitin synthase I plays a major role in cell wall biogenesis of *Neurospora crassa.* Genes Dev 5:2420–2430

Analysis of Plant-Substratum Adhesives

VALERIE VREELAND and LYNN EPSTEIN

1 Introduction

1.1 Principles of Adhesion

"Adhesives are social substances. They unite materials, creating a whole that is greater than the sum of its parts" (Skeist and Miron 1990). The principles of industrial adhesives are equally applicable to plant adhesives. Adhesive bonding is studied with a multidisciplinary approach, including aspects of surface chemistry, rheology, wetting, mechanical properties and fracture mechanics (Pocius 1986). The biological dimension provides additional levels of complexity in plant adhesive formation.

Adhesion is both a process and the interaction which results from the process (Gent and Hamed 1990). Adhesion is defined as "the process by which two surface join to each other" as well as "the sticking together of surfaces or parts" (Morris 1992). For adhesion to occur, either one or both surfaces are sticky, or a thin layer of an initially fluid adhesive material sets (Skeist and Miron 1990). The same kinds of forces cause both cohesion (the binding together of similar substances), and adhesion (the joining of two different materials at their surfaces) (Allen 1987). These forces include covalent chemical bonds and noncovalent molecular interactions. The existence and significance of covalent chemical bonding in adhesion have received recent attention (Lee 1991; Allen 1993) and chemical coupling agents improve adhesive properties (Pocius 1986; Gent and Hamed 1990). Noncovalent forces involved in adhesion include van der Waals bonds, hydrogen bonds, hydrophobic interactions and ionic bonds. Interfacial forces can be characterized as electron donor-acceptor (Lewis acid-base) interactions between adhesive and substrate molecules (Gent and Hamed 1990; Lee 1991). These noncovalent bonds typically require close surface proximity and multiple, low-affinity interactions between molecules. As in the case of antigen-antibody and enzyme-substrate interactions, molecular interactions between adhesive molecules are most effective when the surface shapes and charge properties are complimentary. Allen (1987) concluded that the strength of the adhesive bond typically comes from a combination of forces, and the relative importance of each varies from one adhesive to another.

Modern adhesives are polymer-based composites. The adhesive process involves a series of interconnecting mechanisms (Eagland 1988): the adhesive is always liquid during application; spreads on the surfaces to be joined; establishes

Modern Methods of Plant Analysis, Vol. 17
Plant Cell Wall Analysis
Edited by H.-F. Linskens and J.F. Jackson

intermolecular interactions with the substratum; and develops mechanical strength by chemical cross-linking, polymerization, or a change in physical state, referred to as "curing", "setting" or "hardening". The elastic, brittle and durability properties of the set glue depend on the nature of its components and whether it includes a second phase, either rigid or flexible.

1.2 Cell-Substratum Adhesion

This section is intended as an introduction to some recent ideas on plant cell-substratum adhesive formation and not as a comprehensive review. We emphasize examples of "organism-substratum" adhesion in which a plant, alga or fungus produces a relatively nonspecific adhesive substance and adheres to either an abiotic surface, such as a rock in a tide pool, or to the external surface of a plant. In contrast to many cases of cell-cell adhesion where an "adhesin" on one cell binds to a specific receptor on another cell, organism-substratum adhesion is mediated by a glue. Consequently, examples of organism-substratum adhesion can often be studied in the laboratory with model substrata such as plastic or glass.

No adhesion mechanisms or roles of adhesion molecules in a plant system are fully understood at this time. Numerous candidate adhesive components have been proposed in a variety of plant systems, but their actual involvement in the adhesion process has been documented for very few. Putative plant adhesion molecules include proteins, glycoproteins, proteoglycans, carbohydrates and phenolic compounds. It should be remembered that current molecular phylogenies consider higher plants, algae and fungi to be extremely diverse (Taylor et al. 1993), and therefore are likely to differ in adhesion mechanisms and adhesive components. Several adhesive mechanisms have been proposed for diverse types of insects; similar adhesion systems in different groups are attributed to convergent evolution (Walker 1993).

Awareness of the dynamic nature of the structured polymer network surrounding both plants and animal cells is increasing (Bolwell 1993). We will use the term "exocellular matrix" (ECM), recently proposed by Wyatt and Carpita (1993) to emphasize the active cellular control of this essential extracellular compartment of plants. Aspects of plant exocellular matrix biology have been reviewed recently by Bolwell (1993), Carpita and Gibeaut (1993), Cosgrove (1993), Roberts (1989) and Wyatt and Carpita (1993).

Cell-substratum adhesion differs from both plasmalemma-plasmalemma and plasmalemma-ECM adhesion in specificity and location. Cell-cell and cell-ECM adhesion are mediated by specific interactions at the plasmalemma, while plant cell-substratum adhesion is nonspecific and is distant from the cell surface by the thickness of the wall; plant cell adhesion occurs via the fibrillar wall which surrounds plant cells. Knowledge of mechanisms and molecules involved in cell-cell and cell-ECM adhesion in plants (Knox 1992; Zhu et al. 1994; Wyatt and Carpita, 1993) is less advanced than in animal systems (Rzpecki and Waite 1991; Lasky 1992; Luna and Hitt 1992; Feizi 1993; Tanzer et al. 1993; Walker 1993). Prominent features of animal adhesion include cell adhesion to any surface bearing a ligand for which the cell expresses a receptor (Hook et al. 1988) and specific binding of proteins to carbohydrates. While many molecules and exocellular processes could be modified to participate in an adhesion cascade, the most common process in cell-substratum adhesion in the animal kingdom is production of scleroproteins by quinone-tanning (Waite 1990; Rzepecki and Waite 1991). The marine mussel *Mytilus* (Waite 1990) provides perhaps the best-characterized example of animal-substratum adhesion; here, an adhesive phenolic (L-DOPA rich) protein is polymerized exocellularly by oxidases. In plants, ECM-ECM adhesion and cell-substratum adhesion are both mediated by exocellular components located at the surface of the ECM distal to the plasmalemma, and both incorporate ECM fibers into adhesion phenomena by relatively nonspecific interactions. Since plants have adaptable, exocellular phenolic compound-oxidase systems (Leatham et al. 1980; Peter 1989), plants may have adhesive mechanisms similar to the phenolic proteins in sessile marine animals.

Like most aspects of ECM physiology (Bolwell 1993), cell-substratum adhesion involves modification of the existing ECM framework. Wall components could function as the framework of the adhesive, providing anchor sites for adhesion molecules. The simplest mechanism of adhesion is secretion of a sticky molecule which binds both to ECM fibers and to the substratum at their interface. However, a sticky molecule secreted by the plant cell could not migrate readily through the plant ECM, which is usually at least 100 nm thick (Roberts 1989). Plant ECM-substratum adhesion mechanisms appear to require further complexity in order to control stickiness of adhesion molecules during transport through the ECM. Controlled secretion of precursor molecules could be followed by unmasking or by activation of stickiness, or by initiation of chemical reactivity. The process of developing adhesiveness may require several components, none of which are sticky alone. Specific targeting of adhesive components to the ECM and possible enzymatic activities should be considered. Targeting of protein to wall polymer sites could include lectin binding and other specific protein-carbohydrate interactions such as heparin-fibronectin/heparin-vitronectin binding. In addition to quinone-tanning, other possible steps in an adhesion cascade could include exocellular conformational changes, proteolysis, dephosphorylation, or desulfation. The resulting adhesive would consist of a network of ECM polymers carrying adhesive sites generated in situ. Such a localized process could result in adherence by covalent bonding and/or multiple noncovalent interactions. Therefore, the nature of "adhesion molecules" is expected to be diverse, even in a single plant adhesion system. Consequently, the role of individual components in a multicomponent, exocellular, adhesive-generating system can be difficult to evaluate when individual chemical, enzymatic and histochemical methods are used.

Like all other biological processes, the development of adhesiveness is ultimately controlled by gene expression. Plant adhesive formation has the added complication that intracellular control of formation of exocellular sticky sites at the appropriate time and place is indirect, possibly involving secretion and targeting of ECM proteins, and interactions with indirect gene products such as carbohydrates and polyphenols.

1.2.1 Higher Plant-Substratum Adhesion

There are numerous examples of plant species which produce seeds and fruits (Paquet et al. 1986; Gedalovich et al. 1988, 1989), roots (Fahn 1990), and tendrils (Endress and Thomson 1976) which adhere nonspecifically onto essentially inert substrata. Firm attachment between parasitic seed plants and their hosts is a prerequisite for pathogenicity (Riopel and Baird 1987); while *Agalinis purpurea* and *Striga* require specific extracellular compounds from the host to stimulate germination and holdfast initiation, holdfast attachment is nonspecific and occurs on plastic. While there are specific interactions between pollen and stigma, adhesion of pollen grains (Chaloner 1986; Lolle and Cheung 1993) and pollen germ tubes (Sanders et al. 1991) are relatively nonspecific.

Although plant species produce several sticky compounds such as gums, resins, latex, and kino in secretory ducts and lactifers (Fahn 1990), it is unknown whether these compounds are involved in organism-substratum adhesion. These sticky materials are apparently involved in plant defense after wounding or pathogen attack. These materials have some of the expected properties of an organism-substratum glue. That is, the materials spread as a liquid over the plant surface and then solidify, at least partially. Other examples of plant-produced sticky compounds include the material secreted by specialized trichomes called colleters and the mucilage secreted by carnivorous plants in the genera *Drosera* and *Pinguicula* (Fahn 1990).

1.2.2 Adhesion of Fungal Phytopathogens to a Plant Substratum

Most species of plant pathogenic fungi adhere to their plant hosts before penetration (Nicholson and Epstein 1991). Depending upon the fungal species, spores and/or germlings may be adherent. Adhesion is such a common feature amongst infection structures such as appressoria that these cells are defined as being adherent. Adhesion is involved in the maintenance of the pathogen in its infection court (see Nicholson and Epstein 1991), stimulation of the development of oomycete zoospores into cysts (Hardham and Suzaki 1986), recognition of host surface topography by bean rust germlings (Epstein et al. 1987), stimulation of the host resistance response (Smith and Cruickshank 1987), and the ability of appressoria to mechanically penetrate a plant surface (Howard and Ferrari 1989). Fungal adhesives are apparently secreted in a soluble form through its cell wall. Once the material is exocellular, it becomes insoluble; the insolubilization is essential since most fungi only germinate in the presence of water. Consequently, the ability to adhere underwater is common amongst both aquatic (Hyde et al. 1993) and terrestrial fungi.

Adhesion of phytopathogenic fungi is also common amongst both fungi which penetrate aerial surfaces of plants (Hamer et al. 1988) as well as roots (Gubler and Hardham 1988; Gubler et al. 1989). Aerial plant surfaces are coated with hydrophobic polymers; consequently, these surfaces are relatively inert and there is no evidence for specific chemical receptors on leaves, shoots or fruits. In contrast, root surfaces are more chemically complex and fungal-plant root attachment may involve receptors on the root (see Manocha and Chen 1990). Binding of fungi onto roots can occur via fungal lectin-plant carbohydrate (particularly fucosyl) interactions (Hinche and Clarke 1980). However, many root-infecting phytopathogenic fungi also adhere to glass and plastic. For example, zoospores of the oomycete *Phytophthora cinnamomi* exocytose an adhesive protein during the encystment process (Gubler and Hardham 1988). The actual adhesive phase is very short (Sing and Bartnicki-Garcia 1972; Gubler et al. 1989) and occurs on roots and model substrata.

The majority of the fungal adhesive compounds which have been partially characterized are apparently proteins or glycoproteins (Borrebaeck et al. 1984; Epstein et al. 1987; Gubler and Hardham 1988; Tronchin et al. 1988; Estrada-Garcia et al. 1990; Fukayama and Calderone 1991; Tunlid et al. 1992; Kwon and Epstein 1993). Carbohydrates, especially galactosaminoglycans (Pringle 1981), may serve as adhesive compounds, particularly in adhesion of hyphae onto hydrophilic surfaces such as glass.

1.2.3 Algal-Substratum Adhesion

Algae adhere in aquatic and marine environments. Adhesive components are incorporated into the ECM before being lost in the aqueous environment; stickiness is developed underwater. Algal spores, zygotes, rhizoids and holdfasts adhere strongly to a variety of surfaces, but usually more strongly to higher-energy (i.e., hydrophilic) surfaces (Evans 1989; Fletcher and Callow 1992). Algal adhesion was investigated by light and electron microscopy, histochemistry combined with enzymatic extraction, and chemical and enzymatic perturbation.

Most reports of algal adhesion describe a carbohydrate-containing surface mucilage (Evans 1989; Fletcher and Callow 1992; Vreeland et al. 1993a). Acidic carbohydrates are thought to be involved in marine algal adhesion due to their location in the outermost layer of the rhizoid and holdfast ECM. For example, sessile diatoms produce exocellular acidic polysaccharides, mucilages which rarely contain protein (Daniel et al. 1987; Cooksey 1992; Hoagland et al. 1993). Adherent bacteria also produce exocellular mucilage containing a variety of acidic carbohydrates which are thought to be involved in adhesion (Fletcher et al. 1991). Glycoproteins are thought to participate in adhesion in the green alga *Enteromorpha* (Christie et al. 1970; Callow and Evans 1974), the brown algae *Ectocarpus* (Baker and Evans 1973), *Hormoseira* (Forbes and Hallam 1979), *Laminaria* and *Nereocystis* (Oliveira et al. 1980), and in the red alga, *Ceramium* (Chamberlain and Evans 1981). However, no adhesive compounds were isolated in these algal studies and it was not determined whether both carbohydrate and protein, or glycoprotein or proteoglycan were actually responsible for adhesion. Koutoulis et al. (1993) prepared a monoclonal antibody which binds to a scale-associated protein from the unicellular phytoflagellate *Apedinella radians*. This protein was thought to be involved in adhesion of the scales to the cell surface due to its location.

A fiber-phenolic-catalyst mechanism has been proposed for initial adhesion of by the 4-h-old zygote of the brown alga *Fucus* to the substratum (Vreeland et al. 1992a,b, 1993a; Sect. 2.1). In addition, sulfated fucans (Crayton et al. 1974) and the protein vitronectin were implicated in *Fucus* zygote adhesion (Wagner et al. 1992). *Fucus* zygote adhesion is clearly a complex process incorporating multiple components.

2 Case Studies of Cell-Substratum Adhesion

2.1 Substrate Adhesion by the *Fucus* Zygote

Zygote adhesion to rocks is essential for survival of *Fucus* in the intertidal zone where wave force is a major factor. Substrate adhesion by the *Fucus* zygote wall surface follows a developmentally regulated program even in artificial seawater in the absence of nutrients or light (Quatrano 1980). A wall is assembled immediately after fertilization (see Vreeland and Laetsch 1988) and its outer surface becomes sticky approximately 4 h later (Vreeland et al. 1993b). A second adhesion process begins with the secretion of localized, sticky mucilage in 10–12 h zygotes, followed by germination of a tip-growing rhizoid which is surrounded by a thick, sticky mucilage layer.

The zygote adheres nonspecifically to glass, polyethylene, paraffin and Teflon surfaces (Vreeland et al. 1993b). This observation shows that both hydrophilic and hydrophobic interactions are involved in adhesion. Zygotes adhere earlier to glass than to plastic, and are more easily dislodged from Teflon than from glass. Adhesion to Teflon implies that some adhesive interactions are likely to be noncovalent.

The identification of sticky sites on the zygote surface was revealed using uniformly sized, colored polystyrene microspheres (Vreeland et al. 1993b). Adhesive material was first seen on the zygote surface as isolated spots and patches, with additional strands extending from the surface. This pattern implied that the adhesive material was released from vesicles. Adhesive patches gradually fused, and the hemisphere in contact with the substratum was covered with adhesive at 4–6 h. The age of zygotes initiating adhesive formation varied from 2- to 6-h-old within a synchronously fertilized population (Vreeland et al. 1993b). The entire surface of the zygote was sticky after rhizoid germination. Microspheres clumped together after incubation in spent culture medium, revealing that materials were secreted from zygotes which did not become sticky until after release from the wall.

Vreeland et al. (1992a,b, 1993a) proposed a fiber-phenolic-catalyst mechanism of exocellular adhesive formation in which carbohydrate polymers (alginate and fucans) serve as the fibrous framework. Phloroglucinol-based polyphenols are activated at least indirectly by vanadate-requiring bromoperoxidase. The activated (possibly as phenoxy radicals) polyphenols bind nonspecifically to wall polymers, forming both cross-links and multiple, low affinity adhesion sites. Covalent cross-linking of phenols with the substrate is also possible.

Histochemical methods enabled temporal and spatial colocalization of carbohydrate, polyphenol and vanadate peroxidase with the development of adhesion (Vreeland et al. 1992a,b, 1993a). At the time of initial adhesion, sulfated carbohydrate and polyphenol (stained with Toluidine Blue and Fast Red GG, respectively) were released into seawater. Polyphenols were also stained in the wall at the times of initial adhesion and germination. Fast Blue B salt, another polyphenol stain, was especially useful because it did not penetrate the plasmalemma and stained only newly released polyphenols; the stain revealed vesicle secretion of polyphenol. Vanadate peroxidase substrates which form insoluble products were used to localize enzyme activity. O-dianisidine did not penetrate the plasmalemma and revealed vesicular secretion into the wall between the rhizoid and the glass substrate. N,N-dimethyl-p-phenylene-diamine (DEPDA) (Conyers and Kidwell 1991) penetrated the plasmalemma and identified peroxidase secretion at the rhizoid tip. Using DEPDA, enzyme activity was seen in cortical vesicles at 3 h and in a greater number of peroxidase vesicles in the rhizoid cell than in the thallus cell of two-cell embryos at 24 h.

Exocellular peroxidase activity was seen as flattened vesicles at the rhizoid tip, but exocellular activity was masked by the high concentrations of reducing compounds (i.e., polyphenols) secreted by the zygotes. Exogenous vanadate peroxidase added to zygotes before initial adhesion resulted in the production of greatly enhanced quantities of adhesive material (Vreeland et al. 1993a). Masses of adhesive material extended from single sites on the zygote surface, supporting the contentions that secretion of adhesive precursors was localized and that vanadate peroxidase was involved in activation of the adhesive process. Exogenous vanadate peroxidase catalyzed the development of larger masses of adhesive from secreted precursors at lower zygote densities, demonstrating an effect of enzyme-substrate proportions. Exogenous horseradish peroxidase and mushroom polyphenol oxidase also enhanced the formation of zygote-bound adhesive material.

Attempts to compare adhesiveness statistically in experimental treatments by counting bound microspheres were frustrated by the variability in onset of adhesion. Artificial seawater-grown zygote preparations varying in vanadate, bromide, hydrogen peroxide and phloroglucinol appeared to have the expected effects visually, but only a high (1 mM) hydrogen peroxidase concentration consistently inhibited adhesiveness statistically. Addition of bromide or phloroglucinol appeared to increase rhizoid mucilage cross-linking, resulting in a tighter mucilage appearance. Added phloroglucinol had the additional effect of causing a large adhesive halo around zygotes during initial adhesion and germination which were grown on coverslips and then incubated with microspheres in situ.

A nongelling role for alginate in the adhesive mucilage was demonstrated by immunofluorescence and labeling with a gelling probe (Vreeland and Laetsch 1990). Polyguluronate epitopes were localized to the mucilage with monoclonal antibodies but did not participate in gelling interactions as determined with the fluorescent probe. Calcium-mediated interactions such as alginate gelation were seen at the rhizoid tip by loss of adherent microspheres in calcium chelators such as citrate. A role for sulfated fucans in adhesion was indicated by Crayton et al. (1974), who found that zygotes which were grown in sulfate-free artificial seawater supplemented with methionine were nonadherent and sulfated fucans were not localized. However, secretion of other exocellular components such as sulfated phenolic compounds could also be reduced in the absence of sulfate. Immunogold ultrastructural localization of alginate and fucan antigens showed a uniform distribution of wall carbohydrate antigens throughout the mucilage layer, demonstrating localization at the site of adhesion.

At the time of initial adhesion and during rhizoid germination, concentrations of low molecular weight polyphenols in zygotes decreased (Grotkopp 1990; Vreeland et al. 1992, 1993), supporting polyphenol secretion and polymerization during the adhesion process. Similarly, production of putative polyphenols in perinuclear osmiophilic vesicles was seen by TEM in 1 h but not 4.5 h zygotes. The role of polyphenols as adhesion sites was supported by brief staining of zygotes with Fast Red GG, which prevented microsphere binding in the adhesion assay. In vitro reconstitution experiments with extracted polyphenols and purchased oxidase demonstrated that oxidases can catalyze the formation of large, sticky masses from relatively nonadherent polyphenols (Vreeland et al., unpubl.).

A vitronectin-like protein may be involved in zygote adhesion (Wagner et al. 1992). Vitronectin is thought to participate in transmembrane bridge formation with the cytoskeleton of the zygote, because monospecific (but divalent) antivitronectin antibody inhibited adhesion (Quatrano et al. 1991). However, Springer and Haywood-Reid (1993) showed that inhibition of adhesion by divalent antibody can be an artifact of cytoskeletal cross-linking in *Dictyostelium*. Inhibition of adhesion by antivitronectin antibody needs to be confirmed with monovalent antibody fragments. A possible role for vitronectin in linking sulfated fucans to vanadate peroxidase is seen by comparison with ovoperoxidase positioning in the sea urchin vitelline membrane (Larabell and Chandler 1991), where proteoliasin, another integrin-binding protein, binds to ovoperoxidase.

Recent work on extraction and purification of vanadate peroxidase showed that interaction of this enzyme and polyphenols during extraction caused artifactual interactions. Temporary inactivation of the enzyme in addition to rapid removal of polyphenols greatly reduced the intensity of most bands of enzyme activity on polyacrylamide gels. Rogers and Loveless (1985) described another example of artifactual cross-linking by polyphenols, earlier attributed to lectin-like "haemagglutinins" in extracts of *Fucus* and other brown algae. Oxidase reactions and phenolic compounds are known to cause similar problems in other plant systems (Leatham et al. 1980), where they also may be candidate precursors in adhesive interactions.

In conclusion, the adhesive process in *Fucus* zygotes involves controlled secretion of polyphenols, an exocellular function for a unique vanadate peroxidase, and wall carbohydrates. This process is complex because the two classes of secondary gene products (polyphenols and carbohydrates) are variable in structure. The process is also wasteful because not all of the secreted precursors are rapidly bound to the wall. Opportunities for cellular control exist in the timing and location of component secretion, and in localizing the enzyme in the wall. Goodner and Quatrano (1993) proposed a model for molecular interactions in the transmembrane bridge at the rhizoid tip which includes a membrane-bound integrin-like molecule associated with an exocellular vitronectin-like molecule, which in turn is bound to F2, a sulfated fucan component. Initial adhesion in *Fucus* zygotes could include similar membrane-associated interactions functioning to position the fiber-phenolic-peroxidase adhesive components at the ECM surface.

2.2 Adhesion of Conidiospores of the Plant Pathogenic Fungus, *Nectria haematococca*

In *N. haematococca* macroconidia mating population I (anamorph, *Fusarium solani* f. sp. *cucurbitae* race I), adhesion is a developmentally regulated trait that is expressed on a localized region of the spore. The macroconidia are initially nonadherent, but become adherent after exposure to plant host (zucchini fruit) extracts at temperatures >4 °C (Jones and Epstein 1989). The acquisition of adhesion competence occurs after a 10-min incubation. Thus, the phenotypic change from adhesion-incompetent to competent occurs virtually synchronously in a spore population. In addition, it is the first observable change in macroconidia incubated in a germination-inducing medium; germ tube emergence occurs hours later.

Initially, *N. haematococca* macroconidia adhere at either of the spore apices (Kwon and Epstein 1993). At the same time, the adhesion-competent macroconidia produce a "mucilage" at the spore apex which is labeled with the lectin Concanavalin A (Con A) and produce a macroconidial-specific 90-kDa glycoprotein which is labeled with Con A. Spores incubated in water, dilute V8 juice or a defined medium do not become adhesive, do not produce the macroconidial tip mucilage, and do not produce a 90-kDa glycoprotein; however, spores incubated in dilute V8 juice or the defined medium germinate as efficiently as spores incubated in media that induce adhesion competence and germ tube emergence (Kwon and Epstein 1993). Thus, the production of macroconidial tip mucilage at the site of attachment and the production of the 90-kDa glycoprotein are specifically associated with adhesion and are not generally associated with the process of germination. Furthermore, Concanavalin A prevents adhesion, but not germination, of spores. Thus, the same compound which labels the material at the site of adhesion also blocks adhesion. Neither the development of adhesion competence, the production of macroconidial tip mucilage, nor the production of the 90-kDa glycoprotein requires contact with a substratum. Indeed, once the macroconidia are adhesion-competent, the actual physical attachment to a substratum is a relatively temperature-independent process (Jones and Epstein 1989). In this fungus, cell-substratum adhesion is nonspecific; the fungus adheres to polystyrene, glass, plant host and nonhost surfaces.

To isolate adhesion-reduced mutants, Jones and Epstein (1990) chemically mutagenized microconidia, which are haploid and uninucleate spores. After the mutagenized microconidia were cultured to produce macroconidia, the isolates were repeatedly enriched for the nonadhesive phenotype on polystyrene. Two independently derived mutants were selected with macroconidia which adhered less efficiently (Att^-) to both polystyrene and to the hydrophobic plant surface. Interestingly, both adhesion-reduced mutants are temperature-sensitive (Epstein et al. 1994). A classical genetic analysis of progeny derived from one of the mutants (LE1) indicates that a backcrossed strain has a mutation in a single genetic locus, named *Att*1. Electrophoretic analysis of proteins and glycoproteins in a cell surface-enriched preparation indicated that the mutants are not pleiotropically affected. Caesar-TonThat and Epstein (1991) examined the surfaces of the two mutants and the wild type using freeze-substituted cells and transmission electron microscopy. While the macroconidial walls of all three strains were indistinguishable at the nonadherent, "0 hour" time point, after this time, the outer portion of the walls of both mutants was altered, in different ways, around the entire spore surface. The electron micrographs suggested that the adhesion-reduced mutants might overproduce an electron-dense compound which interferes with adhesion rather than not produce the adhesive compound.

While neither the function of the 90-kDa glycoprotein nor the biochemical basis for adhesion reduction in the mutants is yet known, *Nectria haematococca* appears to be a useful model system for studying cell-substratum adhesion. First, the ascomycetous fungus is generally amenable to the same mutational and genetic analyses as *Saccharomyces cerevisae* and *Neurospora crassa*. For example, since the fungus is haploid, mutants can be selected from an f_1 population. Second, the fungus is easily cultured, stored, and manipulated in the laboratory. Third, the population of fungal spores undergoes a rapid and relatively synchronous phenotypic change from nonadhesive to adhesive; this process can be specifically induced after incubation in an adhesion-inducing plant extract. Information gained in this relatively simple system may provide useful information concerning adhesion onto more complex substrata or adhesion of less experimentally manipulable organisms such as rust fungi (Epstein et al. 1987).

3 Methods for the Analysis of Cell-Substratum Adhesives

3.1 Adhesion Systems and Assays

An ideal plant adhesion system would have the following properties: short culture period, large quantities of readily prepared nonadhesive cells, an inducible initiation of adhesion and synchronous adhesive formation. Initiation of adhesion can be regulated either developmentally, hormonally or environmentally. Environmental switches could include light, temperature, nutrients, chemicals and presence of the host organism.

There are two basic tupes of adhesion assays: aggregation assays and binding assays (Hoffman 1992). Binding assays include particle binding to the plant surface and plant binding to the substratum. Flat model surfaces such as plastic or glass can often be used as substrata for adhesion assays, since cell-substratum adhesion is generally relatively nonspecific. In some cases, organisms which adhere to the hydrophobic, aerial surfaces of plants adhere to a greater extent onto hydrophobic surfaces such as polystyrene than onto hydrophilic surfaces such as glass (Sela-Buurlage et al. 1991).

In perhaps the most common quantitative assay for adhesion, a known number of organisms are incubated on a model substratum. After the substratum is washed, the number of either nonadherent or adherent organisms are quantified. With microscopic organisms, the percentage of adherent organisms can be determined with a particle counter (Young and Kauss 1984; Kwon and Epstein 1993), a hemocytometer (Sela-Buurlage et al. 1991) or microscopically (Epstein et al. 1985). Percentage adhesion can also be determined using radioactively labeled organisms and a scintillation counter (Jones and Epstein 1989). In a similar assay, surface adhesiveness can be quantified by incubating uniformly sized polystyrene or glass microspheres with the test organism; after removing the nonadherent spheres, the number of microspheres per organism can be assessed microscopically (Hazen and LeMelle 1990; Vreeland et al. 1993b) or adherent microspheres can be quantified after solvent extraction of pigment from dyed microspheres. Utilization of microspheres as surface-binding particles has the advantage of identifying the spatial and temporal pattern of adhesiveness during adhesive formation (Vreeland et al. 1993b).

The strength of plant-substratum adhesion can be assessed in several ways: the force of fluid needed to dislodge cells from the substratum can be quantitated (Charters et al. 1973; Christie et al. 1973; Hamer et al. 1988); cytocentrifugation can be used to dislodge *Fucus* zygotes from glass surfaces (Wagner et al. 1992); adherent cells can be dislodged with a micromanipulator-driven rod and the requisite force can be calculated according to Adams et al. (1990). The tensile and torsional strengths of an adhesive compound between two surfaces can also be measured mechanically (Vincent 1992). The wettability or physical interactions between an adhesive compound and a surface can be calculated for a drop of adhesive compound on a substrate from measurement of the contact angle of the drop (Naldrett 1992); the more the adhesive spreads on the surface, the lower the contact angle.

3.2 Identification of Adhesive Components

3.2.1 Isolation of Adhesion Mutants

Mutants with reduced adhesiveness have been isolated in the yeast *Candida albicans* and have helped to identify an adhesive glycoprotein (Calderone and

Wadsworth 1988; Saxena et al. 1990). To obtain adhesion-reduced mutants in *Nectria haematococca*, microconidia, which are single-celled, uninucleate and haploid, were treated with the mutagen N-methyl-N′-nitro-N-nitrosoguanidine (NTG) (Jones and Epstein 1990). After the mutagenized microconidia were cultured to produce macroconidia, the isolates were repeatedly enriched for the nonadhesive phenotype on polystyrene. Two independently selected isolates with reduced adherence onto polystyrene and onto plant host surface were recovered. Using a similar enrichment procedure, Hamer et al. (1989) isolated "Smo" mutants of *Magnaporthe grisea*. However, these mutants have abnormal conidial morphology, aberrations in appressorium formation and other developmental abnormalities. Mutants with such large pleiotropic effects will probably not be useful for studies on adhesion. In addition to mutants with gross pleiotropic effects, there are other types of adhesion-reduced mutants which may not assist in identification of the adhesive compound. For example, adhesion-reduced mutants of the bacterium *Pseudomonas fluorescens* did not adhere as well as the wild type apparently because the mutants overproduce a compound which interferes with adhesion (Pringle et al. 1983; Pringle and Fletcher 1986). In addition, an adhesion-modified phenotype may actually be an indirect effect of a more basic cellular mutant (Knox 1992) or even a defect in the fibrillar framework of the exocellular matrix.

After an adhesion-reduced mutant is obtained, in theory, the researcher can identify the adhesion-related compound by either direct biochemical or reverse genetic approaches. A possible disadvantage for the reverse genetic approach is that the carbohydrate and/or phenolic glue components are not direct gene products. Even proteinaceous glues are postranslationally modified, since they are typically glycosylated. Biochemical characterization of exocellular material is often complex because carbohydrates, polyphenols, and even structural proteins vary in size, number of subunits, and linkage types and arrangements. An advantage of a biochemical in comparison to a reverse genetic approach is that isolation of a structural compound on the cell surface should (in theory) be straightforward.

3.2.2 Correlation of Temporal/Spatial Development with Adhesion

Since adhesion is a developmentally regulated trait in many systems, identification of either factors which regulate the development of adhesiveness (Kwon and Epstein 1993) or specific cells or exocellular regions associated with adherence (Hamer et al. 1988; Hardham and Gubler 1990; Rzepecki and Waite 1991; Koutoulis et al. 1993) can be useful in ultimately identifying the adhesive material. For example, the oomycete *Phytophthora cinnamomi* develops from a motile zoospore into an adherent cyst (Hardham and Suzaki 1986). The developmental sequence is accompanied by distinct ultrastructural changes on the cell surface. Using fluorescence and transmission electron microscopy, Hardham and colleagues (Hardham and Suzaki 1990; Hardham et al. 1991) labeled de novo cell surface compounds with monoclonal antibodies and lectins. Their results suggested that a ca. 200-kDa protein is involved in cell-substratum adhesion. Similarly, histochemical stains (Fink 1986; Baker and Evans 1973; Callow and Evans 1974; Forbes and Hallam 1979; Oliviera et al. 1980; Chamberlain and Evans 1981) can be used to correlate the appearance of particular compounds with the development of adhesiveness (Sect. 2.1; Chubal et al. 1991). The sensitivity of histochemical detection can be increased by employing fluorescent stains (Molecular Probes, Inc., Eugene, Oregon). However, negative results with particular stains do not necessarily demonstrate that a class of compounds is not involved in adhesion; adhesive molecules are highly modified and, consequently, the standard reactive sites for histochemical reactions may be blocked.

Since a cell-substratum glue must be present on the cell surface, cell surface labeling can be an effective method for tagging adhesive compounds. After extraction, putative adhesive molecules can be identified by comparing the cell surface compounds (1) before and after the development of adhesiveness or (2) on adhesion mutants vs. wild type. Cell-surface proteins can be covalently labeled with ^{125}I (Borrebaeck et al. 1984; Epstein et al. 1987) or sulfo-NHS-biotin (Sargiacomo et al. 1989; Kwon and Epstein 1992). Cell surface carbohydrates can be covalently labeled with biotin hydrazide (Lisanti et al. 1989).

Relatively new methods are available for direct chemical characterization of polymers adsorbed onto solid surfaces. For example, using Fourier transform infrared spectroscopy, Tunlid et al. (1991) provided evidence that encysting zoospores of *Catenaria anguillulae* adhere onto a germanium crystal via protein(s).

Two general precautions should be taken in interpreting the correlation between the development of adhesiveness and the appearance of a cell surface molecule. First, it must be remembered that ECM and cell surfaces are complex (Nicholson and Epstein 1991; Pascholati et al. 1992) and that the coincidental appearance of a compound and a developmentally regulated trait does not prove that the compound is involved in the particular trait. Second, appropriate controls to detect false positives are critically important in experiments involving antibodies, lectins and histochemical reagents (Springer and Haywood 1993; Wick 1993), since adhesive compounds can bind test materials via nonspecific rather than by the desired specific mechanisms.

3.2.3 Experimental Perturbation of Adhesion

Classes of adhesive compounds have been deduced by monitoring for a decrease in adhesion after treatments which either degrade or block specific chemical moieties. Typical treatments include periodate oxidation to detect carbohydrates with vicinal hydroxyls (Callow and Evans 1974; Hinch and Clarke 1980) and various enzymes which do not enter the cell (Christie et al. 1970; Oliveira et al. 1980; Epstein et al. 1985, 1987). For enzymatic analyses, only relatively pure enzymes should be used. Other treatments which might prevent adhesion include histochemical stains (Sect. 2.1; Chubal et al. 1991), reagents which cleave or modify particular chemical groups (Fry 1988; Tunlid et al. 1992), lectins (Kwon and

Epstein 1993) and antibodies (Springer and Haywood-Reid 1993). In all treatments, controls should be included since pH, specific ions and buffers, and ionic strength can affect adhesion. In experimental perturbation of adhesion, it is also prudent to monitor cell viability, since gross denaturation of the cell or its surface may affect cell adhesion. A limitation of adhesion perturbation experiments is that they do not differentiate between sticky adhesion compounds, molecules necessary for the formation of the glue or materials attached to adhesive compounds.

3.3 Analysis of Adhesive Components

3.3.1 Extraction and Purification

The development of a strategy for purification and analysis of a cell-substratum adhesive compound obviously depends upon its composition and conditions required for solubilization. Solubilization is a particular problem with adhesive compounds since glues are essentially insoluble. Moreover, many botanical glues must function in an aqueous environment, and consequently, must be water-insoluble. The rationale for a water-insoluble glue in an aquatic alga is obvious. Glues produced by phytopathogenic fungi on plant surfaces are also water-insoluble; since most fungal spores require "free water" to germinate, plant surfaces are, at least intermittently, aqueous environments.

To select an appropriate detergent and/or solvent for extraction (Fry 1988; Tunlid et al. 1992), researchers can determine the efficacy of a variety of detergents, chaotropic agents, alkalies, acids, organic solvents, and salts in detaching the adherent organism from its substratum. For example, with the fungus *Nectria haematococca*, spores were detached and the adhesive material was presumably extracted by incubating intact spores at 100 °C for 5–10 min in Laemmli buffer (625 mM Tris-HCl, pH 6.8, 2% SDS, and 5% β-mercaptoethanol) (Kwon and Epstein 1993). However, the detachment assay does not differentiate between agents which (1) actually solubilize the adhesive material, (2) disrupt the bonds at the cell-substratum interface without solubilization of the adhesive material, (3) hydrolyze and then solubilize the partially degraded adhesive material, or (4) either solubilize or denature at least one of the adhesive components of the glue before it hardens into an insoluble matrix.

Extracting soluble precursors of the adhesive material avoids many of the problems associated with purifying insoluble compounds. Presumably, adhesive compounds are secreted onto the cell surface in a soluble form and then rendered insoluble after cross-linking. For example, precursors of the insoluble plant cell wall glycoprotein extensin are readily extracted with 50 mM $CaCl_2$ (Cooper and Varner 1983; Smith et al. 1984); after cross-linking by cell wall peroxidases, extensin is not extractable. Similarly, characterization of enzymes, cofactors, and ions that are involved in the adhesion process could assist in developing a strategy to purify the adhesion precursors. For example, if oxidases (Sect. 2.1; Waite 1990) or Ca^{2+} (Gubler et al. 1989) are involved in cross-linking, appropriate inhibitors or sequestering agents could be added to the extraction medium.

Regardless of whether the adhesive material is a cross-linked polymer, nonspecific cross-linking by oxidases may have to be inhibited in order to successfully purify an adhesive compound. Many higher plants, fungi and algae contain exocellular oxidative enzymes and phenolic compounds; the enzymes can cross-link proteins to themselves or to other materials in the extract (Leatham et al. 1980; Sect. 2.1). For example, unless vanadate peroxidase is inactivated at the start of extraction from *Fucus*, the enzyme apparently becomes cross-linked to sulfated fucans (Sect. 2.1). Some oxidases require specific cations for activity and can be inactivated by chelation. Enzymes can be denatured, for example, by extracting exocellular compounds into Laemmli buffer at 100 °C (Kwon and Epstein 1993). The reactivity of phenolic compounds can be minimized by including in the extraction buffer polyvinylpolypyrollidone (PVPP) and antioxidants (Loomis and Bataille 1966; Loomis 1974; Gegenheimer 1990).

The exocellular adhesive material also could be extracted after specifically cleaving the cross-linked sites. For example, acidified chlorite cleaved extensin, more or less specifically at the isodityrosine cross-links (Biggs and Fry 1990) and phenolic ester or ether linkages to wall polymers can be differentially cleaved (Lam et al. 1992).

Since adhesive compounds are "sticky," they can be readily lost by nonspecific adsorption onto glassware, plasticware or filters. Therefore, the number of preparative steps should be minimized. Concentration of material in dialysis tubing coated with flakes of polyethylene glycol (MW 4000–8000) minimizes contact with membrane filters. Purification of the solubilized adhesive material may require the presence of detergent, chaotrope and reducing agents (Waite and Benedict 1984). For initial purification procedures, SDS-gel electrophoresis may be the method of choice (Kwon and Epstein 1993). Alternatively, nonspecific adsorption could be used in the purification strategy; once the adhesive compounds are extracted, they could be adsorbed onto plastic or glass beads and then washed free of many contaminating compounds before elution with chaotropes or detergents.

Purified adhesion molecules can be analyzed by standard methods for cell walls (Fry 1988), including carbohydrates, proteins (Deutscher 1990), polyphenols (Fry 1988; Harborne 1989; Porter 1989; Yamamoto et al. 1989; Lewis 1993) and phenolic cross-links (Azuma 1989; Fry 1989; Lam et al. 1992; Van Der Hage et al. 1993).

3.3.2 Assays of Isolated Compounds for Adhesive Activity

Once a putative molecule is isolated, monospecific antibodies can be produced and used to determine if: (1) the compound of interest is present in the adhesive region (Moore 1989); and (2) the antibodies disrupt adhesion. However, there are risks of false positives in antibody-disruption experiments. Controls must be carefully selected so that specific labeling can be differentiated from nonspecific adsorption to a sticky molecule. Antibody interactions with cytoskeletal-exocellular matrix bridging may also affect adhesion. Springer and Haywood-Reid (1993) showed that in *Dictyostelium* divalent IgG antibodies to gp40 inhibited adhesion but monovalent Fab fragments of the same antibody did not inhibit adhesion. They concluded that the divalent antibody cross-linked gp40 proteins on the cell surface, causing a cytoskeletal change which prevented cell-cell and cell-substratum adhesion. With antibodies, there are also risks of false negatives; an antibody which binds to a cell-substratum molecule may become a part of the glue rather than inhibit the glue.

After all of the structural components and cross-linking mechanisms of an adhesive system have been identified, the researcher should develop an in vitro reconstitution system and demonstrate its adhesive function (Sect. 2.1). Assays for adhesive strength or stickiness may be thwarted unless glue hardening is controlled by maintaining the adhesive compound in a soluble form.

4 Conclusions

Recent progress has been slow in understanding plant adhesion to the substratum and a thorough understanding of an adhesion process in a plant system is yet to be attained. The search for a single sticky molecule has not been productive in most plant adhesion systems. However, evidence for multiple mechanisms of adhesion is available from investigations on a variety of plant systems. Common themes include sticky glycoproteins, enzymatic oxidation of phenolic compounds into a cross-linked adhesive network, and a role for carbohydrates. The particular molecules involved appear to differ in diverse plant systems. Therefore, detailed methods of detection, extraction and purification must be tailored for each system.

Particular attention to selection of suitable plant systems is needed for identification of adhesive components. Adhesive precursors must be fully characterized and their role in adhesion clearly demonstrated. Although a strong correlation of candidate molecules with adhesion can be achieved, conclusive documentation of the role of that molecule in the adhesion process is difficult but essential. The stickiness of adhesive components requires special consideration to avoid loss during extraction and faulty conclusions during analysis. Development of methods to address plant adhesion studies is essential for future progress. The concept of cell-substratum adhesion as an exocellular matrix process, integrating specific interactions between adhesive precursors and matrix polymers to form a composite glue network, should facilitate progress in understanding plant adhesive mechanisms and components.

Acknowledgments. This work was supported in part by the USDA NRI Competitive Grants Program Grants 9300677 to L. Epstein and 9203132 to J.A. West, the DOE/NSF/USDA Multi-Institutional Research Co-ordination Program Group Grant 92-37310-7821 to L. Epstein et al. and by the NSF Developmental Biology Program Grant DCB8905221 to W.M. Laetsch.

References

Adams DS, Keller R, Koehl MAR (1990) The mechanics of notochord elongation, straightening and stiffening in the embryo of *Xenopus laevis*. Development 110:115–130

Allen KW (1987) A review of contemporary views of theories of adhesion. In: Sharpe LH (ed) Adhesion international, Proc Annu Meet Adhesion Soc 10:85–101

Allen KW (1993) Some reflections on contemporary views of theories of adhesion. Int J Adhesion Adhesives 13:67–72

Azuma JI (1989) Analysis of lignin-carbohydrate complexes of plant cell walls. Linskens HF, Jackson JF (eds) In: Modern methods of plant anaylsis. Springer, Vol. 10. Berlin Heidelberg New York, pp 100–126

Baker JR, Evans LV (1973) The ship-fouling alga *Ectocarpus*. I. Ultrastructure and cytochemistry of plurilocular reprodutive stages. Protoplasma 77:1–13

Biggs KJ, Fry SC (1990) Solubilization of covalently bound extensin from *Capsicum* cell walls. Plant Physiol 92:197–204

Bolwell GP (1993) Dynamic aspects of the plant extracellular matrix. Int J Cytol 146:261–324

Borrebaeck CAK, Mattiasson B, Nordbring-Hertz B (1984) Isolation and partial characterization of a carbohydrate-binding protein from a nematode-trapping fungus. J Bacteriol 159:53–56

Caesar-TonThat TC, Epstein L (1991) Adhesion-reduced mutants and the wild type *Nectria haematococca*: an ultrastructural comparison of the macroconidial walls. Exp Mycol 15:193–205

Calderone RA, Wadsworth E (1988) Characterization of mannoproteins from a virulent *Candida albicans* strain and its derived, avirulent strain. Rev Infect Dis 10:S423–S427

Callow ME, Evans LV (1974) Studies on the ship-fouling alga *Enteromorpha*. III. Cytochemistry and autoradiography of adhesive production. Protoplasma 80:15–27

Carpita NC, Gibeaut DM (1993) Structural models of primary walls in flowering plants: consistence of molecular structure with the physical properties of the walls during growth. Plant J 3:1–30

Chaloner WG (1986) Electrostatic forces in insect pollination and their significance in exine ornament. In: Blackmore S, Ferguson II (eds) Pollen and spores: form and function. Academic Press, London, pp 103–108

Chamberlain AHL, Evans LV (1981) Chemical and histochemical studies of the spore adhesive of *Ceramium*. In: Fogg GE, Jones WE (eds) Proc Int Seaweed Symp 8, University College of North Wales, Menai Bridge, pp 538–542

Charters AC, Neushul M, Coon D (1973) The effect of water motion on algal spore adhesion. Limnel Oceanogr 18:884–896

Christie AO, Evans LV, Shaw M (1970) Studies on the ship-fouling alga *Enteromorpha*. II. The effect of certain enzymes on the adhesion of zoospores. Ann Bot 34:466–482

Chubal R, Wilmot VA, Wynn WK (1991) Visualization, adhesiveness, and cytochemistry of the extracellular matrix produced by urediniospore germ tubes of *Puccinia sorghi*. Can J Bot 69:2044–2054

Conyers SM, Kidwell DA (1991) Chromogenic substrates for horseradish peroxidase. Anal Biochem 192:207–211

Cooksey KE (1992) Extracellular polymers in biofilms. In: Melo LF et al. (eds) Biofilms – science and technology. Kluwer, Dordrecht, pp 137–147

Cooper JB, Varner JE (1983) Solubilization of hydroxyproline-rich cell wall glycoprotein in aerated carrot root slices. Biochem Biophys Res Commun 112:161–167

Cosgrove DJ (1993) Wall extensibility: its nature, measurement and relationship to plant cell growth. New Phytol 124:1–23

Crayton MA, Wilson E, Quatrano RS (1974) Sulfation of fucoidan in *Fucus* embryos. II. Separation from initiation of polar growth. Dev Biol 39:164–167

Daniel GF, Chamberlain AHL, Jones EBG (1987) Cytochemical and electron microscopical observations on the adhesive materials of marine fouling diatoms. Br Phycol J 22:101–118

Deutscher MP (1990) Guide to protein purification. Methods Enzymol 182.

Eagland D (1988) Adhesion and adhesive performance: the scientific background. Endeavor 12:179–184
Endress AG, Thomson WW (1976) Ultrastructural and cytochemical studies on the developing adhesive disc of boston ivy tendrils. Protoplasma 88:315–331
Epstein L, Laccetti L, Staples RC, Hoch HC, Hoose WA (1985) Extracellular proteins associated with induction of differentiation in bean rust uredospore germlings. Phytopathology 75:1073–1076
Epstein L, Laccetti L, Staples RC, Hoch HC (1987) Cell-substratum adhesive proteins involved in surface contact responses of the bean rust fungus. Physiol Mol Plant Pathol 30:373–388
Epstein L, Kwon YH, Almond DE, Schached LM, Jones MJ (1994) Genetic and biochemical characterization of *Nectria haematococca* strains with adhesive and adhesion-reduced macroconidia. Appl Environ Microbiol 60:524–530
Estrada-Garcia MT, Callow JA, Green JR (1990) Monoclonal antibodies to the adhesive cell coat secreted by *Pythium aphanidermatum* zoospores recognise 200×10^3 Mr glycoproteins stored within large peripheral vesicles. J Cell Sci 95:199–206
Evans LV (1989) Mucilaginous substances from macroalgae: an overview. (1989) In: Chantler E, Ratcliffe NA (eds) Symposia of the Society for Experimental Biology MUCUS and Related Topics 43, The Company of Biologists Ltd, Cambridge, pp 455–461
Fahn A (1990) Plant anatomy, 4th edn. Pergamon Press, Oxford
Feizi T (1993) Oligosaccharides that mediate mammalian cell-cell adhesion. Curr Opinion Structural Biol 3:701-710
Fink S (1986) A new integrated concept for the improved preparation of sections of fresh or frozen tissue for light microscope histochemistry. Histochemistry 86:43–52
Fletcher M, Lessmann JM, Loeb GI (1991) Bacterial surface adhesives and biofilm matrix polymers of marine and freshwater bacteria. Biofouling 4:129–140
Fletcher RL, Callow ME (1992) The settlement, attachment and establishment of marine algal spores. Br Phycol J 27:303–329
Forbes MA, Hallam ND (1979) Embryogenesis and substratum adhesion in the brown alga *Hormosira banksii* (Turner) Decaisne. Br Phycol J 14:69–81
Fry SC (1988) The growing plant cell wall: chemical and metabolic analysis. Longman, New York
Fry SC (1989) Analysis of cross-links in the growing cell walls of higher plants. In: Linskens HF, Jackson JF (eds) Modern methods of plant anaylsis, Vol. 10. Springer, Berlin Heidelberg New York, pp 12–36
Fukayama M, Calderone RA (1991) Adherence of cell surface mutants of *Candida albicans* to buccal epithelial cells and analyses of the cell surface proteins of the mutants. Infect Immun 59:1341–1345
Gahan PB (1984) Plant histochemistry and cytochemistry: an introduction. Academic Press, London
Gedalovich E, Kujit J, Carpita NC (1988) Chemical composition of viscin, an adhesive involved in dispersal of the parasite *Phoradendron californicum* (Viscaceae). Physiol Mol Plant Pathol 32:61–76
Gedalovich-Shedletzky E, Delmer DP, Kujit J (1989) Chemical composition of viscin mucilage from three mistletoe - a comparison. Ann Bot 64:249–252
Gegenheimer P (1990) Preparation of extracts from plants. Methods Enzymol 182:174–193
Gent AN, Hamed GR (1990) Fundamentals of adhesion. In: Skeist I (ed) Handbook of adhesives, 3rd edn. Van Nostrand Reinhold, New York, pp 39–73
Glombitza KW, Knoss W (1992) Sulphated phlorotannins from the brown alga *Pleurophycus gardneri*. Phytochemistry 31:279–281
Goodner B, Quatrano RS (1993) *Fucus* embryogenesis: a model to study the establishment of polarity. Plant Cell 5:1471–1481
Grotkopp E (1990) Changes in phenolic content of *Fucus* embryos: implications for its role in early development. Senior Honors Thesis, Department of Integrative Biology, University of California, Berkeley
Gubler F, Hardham AR (1988) Secretion of adhesive material during encystment of *Phytophthora cinnamomi* zoospores, characterized by immunogold labelling with monoclonal antibodies to components of peripheral vesicles. J Cell Sci 90:225–235

Gubler F, Hardham AR, Duniec J (1989) Characterising adhesiveness of *Phytophthora cinnamomi* zoospores during encystment. Protoplasma 149:24–30

Hamer JE, Howard RJ, Chumley FG, Valent B (1988) A mechanism for surface attachment in spores of a pathogenic fungus. Science 239:288–290

Hamer JE, Valent B, Chumley FG (1989) Mutations at the SMO genetic locus affect the shape of diverse cell types in the rice blast fungus. Genetics 122:351-361

Hames BD, Rickwood D (eds) (1990) Gel electrophoresis of proteins: a practical approach, 2nd edn. IRL Press, Oxford

Harborne J (1989) General procedures and measurement of total phenolics. Methods Plant Biochem 1:2–28

Hardham AR, Suzaki E (1986) Encystment of zoospores of the fungus, *Phytophthora cinnamomi*, is induced by specific lectin and monoclonal antibody binding to the cell surface. Protoplasma 133:165–173

Hardham A, Gubler F (1990) Polarity of attachment of zoospores of a root pathogen and pre-alignment of the emerging germ tube. Cell Biol Int Rep 14:947–956

Hardham AR, Suzaki E (1990) Glycoconjugates on the surface of spores of the pathogenic fungus *Phytophthora cinnamomi* studied using fluorescence and electron microscopy and flow cytometry. Can J Microbiol 36:183–192

Hardham AR, Gubler F, Duniec J, Elliott J (1991) A review of methods for the production and use of monoclonal antibodies to study zoosporic plant pathogens. J Microsc 162:305–318

Hazen KC, LeMelle WG (1990) Improved assay for surface hydrophobic avidity of *Candida albicans* cells. Appl Environ Microbiol 56:1974–1976

Hemingway RW, Laks PE (1992) Plant polyphenols, synthesis, properties, significance. Plenum Press, New York

Hinch JM, Clarke AE (1980) Adhesion of fungal zoospores to root surfaces is mediated by carbohydrate determinants of the root slime. Physiol Plant Pathol 16:303–307

Hoagland KD, Rosowski JR, Gretz MR, Roemer SC (1993) Diatom extracellular polymeric substances: function, fine structure, chemistry and physiology. J Phycol 29:537–566

Hoffman S (1992) Assays of cell adhesion. In: Stevenson BR, Galin WJ, Paul DJ (eds) Cell-cell interactions, a practical approach. IRL Press, Oxford, pp 1–30

Hook M, Woods A, LeBaron RG, Switalski LM, Johansson S (1988) Molecular mechanisms of cell-substrate adhesion. In: Varner JE (ed) Self-assembling architecture. Alan R Liss, New York, pp 119-127

Howard RJ, Ferrari MA (1989) The role of melanin in appressorium formation. Exp Mycol 13:403–418

Hyde KD, Greenwood R, Jones EBG (1993) Spore attachment in marine fungi. Mycol Res 97:9–14

Jones MJ, Epstein L (1989) Adhesion of *Nectria haematococca* macroconidia. Physiol Mol Plant Pathol 35:453–461

Jones MJ, Epstein L (1990) Adhesion of macroconidia to the plant surface and virulence of *Nectria haematococca.* Appl Environ Microbiol 56:3772–3778

Knox JP (1992) Cell adhesion, cell separation and plant morphogenesis. Plant J 2:137–41

Knox JP (1993) The role of cell surface glycoproteins in differentiation and morphogenesis. In: Battery NH, Dickinson HG, Hetherington AM (eds) Post-translational modifications in plants. Cambridge University Press, Cambridge, pp 267–283

Knox RB (1984) Pollen-pistil interactions. In: Linskens HF, Heslop-Harrison J (eds) Cellular interactions. Encyclopedia of plant physiology, vol 17. Springer Berlin Heidelberg New York, pp 508–560

Koutoulis A, Ludwig M, Wetherbee R (1993) A scale associated protein of *Apedinella radians* (Pedinellophyceae) and its possible role in the adhesion of surface conmponents. J Cell Sci 104:391–398

Kwon YH, Epstein L (1992) Biotin-labeling identifies glycoproteins in cell walls of *Nectria haematococca* macroconidia. Phytopathology 82:1127 (Abstr)

Kwon YH, Epstein L (1993) A 90 kDa glycoprotein associated with adhesion of *Nectria haematococca* macroconidia to substrata. Mol Plant-Microbe Interactions 6:481–487

Lam TBT, Iiyama K, Stone BA (1992) Cinnamic acid bridges between cell wall polymers in wheat and *Phalaris* internodes. Phytochemistry 31:1179–1183

Larabell C, Chandler DE (1991) Fertilization-induced changes in the vitelline envelope of

echinoderm and amphibian eggs: self-assembly of an extracellular matrix. J Electron Microsc Tech 17:294–318
Lasky L (1992) Selectins: interpreters of cell-specific carbohydrate information during inflammation. Science 258:964–969
Leatham GF, King V, Stahmann MA (1980) In vitro protein polymerization by quinones or free radicals generated by plant or fungal oxidative enzymes. Phytopathology 70:1134–1140
Lee L-H (1991) Recent studies in polymer adhesion mechanisms. In: Lee L-H (ed) Adhesive bonding. Plenum Press, New York, pp 1–30
Lewis NG (1993) Plant phenolics. In: Alscher RG, Hess JL (1993) Antioxidants in higher plants. CRC Press, Boca Raton, pp 135–169
Lisanti MP, Le Bivic A, Sargiacomo M, Rodriguez-Boulan E (1989) Steady-state distribution and biogenesis of endogenous Madin-Darby canine kidney glycoproteins: evidence for intracellular sorting and polarized cell surface delivery. J Cell Biol 109:2117–2127
Lolle SJ, Cheung AY (1993) Promiscuous germination and growth of wildtype pollen from *Arabidopsis* and related species on the shoot of the *Arabidopsis* mutant, fiddlehead. Dev Biol 155:250–258
Loomis WD (1974) Overcoming problems of phenolics and quinones in the isolation of plant enzymes and organelles. Methods Enzymol 31:528–545
Loomis WD, Battaile J (1966) Plant phenolic compounds and the isolation of plant enzymes. Phytochemistry 5:423–438
Luna ET, Hitt AL (1992) Cytoskeleton-plasma membrane interactions. Science 258:955–964
Manocha MS, Chen Y (1990) Specificity of attachment of fungal parasites to their hosts. Can J Microbiol 36:69–76
Moore PJ (1989) Immunogold localization of specific components in plant cell walls. In: Linskens HF, Jackson JF (eds) Modern methods of plant analysis, Vol. 10. Springer, Berlin Heidelberg New York, pp 70–88
Morris C (ed) (1992) Dictionary of science and technology, Academic Press, San Diego, p 43
Naldrett MJ (1992) Adhesives. In: Vincent JFV (ed) Biomechanics – materials, a practical approach. IRL Press, Oxford, pp 219–239
Nicholson RL, Epstein L (1991) Adhesion of fungi to the plant surface: prerequisite for pathogenesis. In: Cole GT, Hoch HC (eds) The fungal spore and disease initiation in plants and animals. Plenum Press, New York, pp 3–23
Oliviera L, Walker DC, Bisalputra T (1980) Ultrastructural, cytochemical, and enzymatic studies on the adhesive "plaques" of the brown alga *Laminaria saccharina* (L.) Lamour, and *Nereocystis luetkeana* (Nert.) Post. er Rupr. Protoplasma 104:1–15
Paquet PJ, Knutson DM, Tinnin RO, Tocher RD (1986) Characteristics of viscin from the seeds of dwarf mistletoe. Bot Gaz 147:156–158
Pascholati SF, Yoshioka H, Kunoh H, Nicholson RL (1992) Preparation of the infection court by *Erysiphe graminis* f. sp. *hordei*: cutinase is a component of the conidial exudate. Physiol Mol Plant Pathol 41:53–59
Peter MG (1989) Chemical modifications of biopolymers by quinones and quinone methides. Angew Chem Int Ed Engl 28:555–570
Pocius AV (1986) Fundamentals of adhesive bonding. In: Hartshorn SR (ed) Structural adhesives, chemistry and technology. Plenum Press, New York, pp 23–68
Porter LJ (1989) Tannins. Methods Plant Biochem 1:389–419
Pringle RB (1981) Nonspecific adhesion of *Bipolaris sorokiniana* sporelings. Can J Plant Pathol 3:9–11
Pringle JH, Fletcher M (1986) Adsorption of bacterial surface polymers to attachment substrata. J Gen Microbiol 132:743–749
Pringle JH, Fletcher M, Ellwood DC (1983) Selection of attachment mutants during the continuous culture of *Pseudomonas fluorescens* and relationship between attachment ability and surface composition. J Gen Microbiol 129:2557–2569
Quatrano RS (1980) Gamete release, fertilization and embryogenesis in the Fucales. In: Gantt E (ed) Handbook of phycological methods. Cambridge University Press, Cambridge, pp 59–68
Quatrano RS, Brian L, Aldridge J, Schultz T (1991) Polar axis fixation in *Fucus* zygotes: components of the cytoskeleton and extracellular matrix. Development (Suppl) 1:11–16
Reid PD, Pont-Lezica R (eds) (1992) Tissue printing. Academic Press, San Diego

Riopel JL, Baird WV (1987) Morphogenesis of the early development of primary haustoria in *Striga asiatica*. In: Musselman LJ (ed) Parasitic weeds in agriculture, vol 1. CRC Press, Boca Raton, pp 91–106
Roberts K (1989) The plant extracellular matrix. Curr Opinion Cell Biol 1:1020–1027
Rogers DJ, Loveless RW (1985) "Haemagglutinins" of the Phaephyceae and nonspecific aggregation phenomena by polyphenols. Bot Mar 28:133–137
Rzepecki LM, Waite HJ (1991) DOPA proteins: versatile varnishes and adhesives from marine fauna. Bioorg Mar Chem 4:120–148
Sanders LC, Lord EM et al. (1991) A homolog of the substrate adhesion molecure vitronectin occursin a species of flowering plant. Plant Cell 3:629–636
Sargiacomo M, Lisanti M, Graeve L, LeBivic A, Rodriguez-Boulan E (1989) Integral and peripheral protein composition of the apical and basolateral membrane domains in MDCK Cells. J Membr Biol 107:277–286
Saxena A, McElhaney-Feser GE, Cihlar RL (1990) Mannan composition of the hyphal form of two relatively avirulent mutants of *Candida albicans*. Infect Immun 58:2061–2066
Sela-Buurlage MB, Epstein L, Rodriguez RJ (1991) Adhesion of ungerminated *Colletotrichum musae* conidia. Physiol Mol Plant Pathol 39:345–352
Sing VO, Bartnicki-Garcia S (1972) Adhesion of zoospores of *Phytophthora palmivora* to solid surfaces. Phytopathology 62:790
Skeist I, Miron J (1990) Introduction to adhesives. In: Skeist I (ed) Handbook of adhesives, 3rd edn. Van Nostrand Reinhold, New York, pp 3–20
Smith JJ, Muldoon EP, Lamport DTA (1984) Isolation of extensin precursors by direct elution of intact tomato cell suspension cultures. Phytochemistry 23:1233–1239
Smith MM, Cruickshank IAM (1987) Dynamics of conidial adhesion and elicitor uptake in relation to pisatin accumulation in aqueous droplets on the endocarp of pea pods. Physiol Mol Plant Pathol 31:315–324
Springer WR, Haywood-Reid PL (1993) Antibodies specific for gp40 inhibit cell-cell adhesion by crosslinking the protein on the surface of *Dictyostelium purpureum*. J Cell Biochem 53:85–97
Tanzer ML, Giniger MS, Dean JW, Chandrasekaran S (1993) The role of glycosylation in adhesive protein function. In: Roberts DD, Mecham RP (eds) Cell surface and extracellular glycoconjugates. Academic Press, San Diego, pp 271–308
Taylor J, Swann E, Berbee ML (1994) Molecular evolution of ascomycete fungi: phylogeny and conflict. In: Hawkesworth DL (ed) 1st Int Worksh on Ascomycete systematics. NATO Adv Sci Inst Ser Plenum press, New York
Tronchin G, Bouchara JP, Robert R, Senet JM (1988) Adherence of *Candida albicans* germ tubes to plastic: ultrastructural and molecular studies of fibrillar adhesins. Infect Immun 56:1987–1993
Tunlid A, Jansson HB, Nordbring-Hertz B (1992) Fungal attachment to nametodes. Mycol Res 96:401–412
Tunlid A, Nivens, DE, Jansson HB, White DC (1991) Infrared monitoring of the adhesion of *Catenaria anguillulae* zoospores to solid surfaces. Exp Mycol 15:206–214
Van Der Hage ERE, Mulder MM, Boon JJ (1993) Structural characterization of lignin polymers by temperature-resolved in-source pyrolysis mass spectrometry and curie-point pyrolysis gas chromatography mass spectrometry. J Anal Appl Pyrolysis 25:77–95
Vincent JFV (1992) Plants. In: Vincent JFV (ed) Biomechanics - materials, a practical approach. IRL Press at Oxford University Press, Oxford, pp 165–191
Vreeland V, Laetsch WM (1988) The role of alginate self-associating subunits in the assembly of *Fucus* embryo cell walls. In: Varner JE (ed) Self-assembling architecture. Alan R Liss, New York, pp 77–96
Vreeland V, Laetsch WM (1990) A gelling carbohydrate in algal cell wall formation. In: Adair WS, Mecham RP (eds) Organization and assembly of plant and animal extracellular matrix. Academic Press, San Diego, pp 137–171
Vreeland V, Wang X, Grotkopp E, Laetsch WM (1990) Non-gelling wall interactions of alginate in *Fucus* zygotes correlate with adhesion. American Society for Cell Biology Summer Conference, Airlie, VA, P20B (Abstr)
Vreeland V, Grotkopp E, Espinosa S, Quiroz D, Laetsch WM, West JA (1992a) Formation of a plant cell wall adhesive. 6th Int Cell wall Meet Abstr, Nijmegen, The Netherlands, p 251

Vreeland V, Grotkopp E, Espinosa S, Quiroz D, West J (1992b) *Fucus* zygote adhesion. 14th Int Seaweed Symp Abstr, p 130

Vreeland V, Grotkopp E, Espinosa S, Quiroz D, Laetsch WM, West JA (1993a) Formation of *Fucus* cell wall adhesive. J Cell Biochem Suppl (17 Part A), p 23 (Abstr)

Vreeland V, Grotkopp E, Espinosa S, Quiroz D, Laetsch WM, West J (1993b) The pattern of cell wall adhesive formation by *Fucus* zygotes. Hydrobiologia 260/261:485–491

Wagner VT, Brian L, Quatrano RS (1992) Role of a vitronectin-like molecule in embryo adhesion of the brown alga *Fucus*. Proc Natl Acad Sci USA 89:3644–3648

Waite JH, Benedict CV (1984) Assay of dihydroxy phenylalanine (Dopa) in invertebrate structural proteins. Methods Enzymol 107:397–413

Waite JH (1990) The phylogeny and chemical diversity of quinone-tanned glues and varnishes. Comp Biochem Physiol 97B:19–29

Walker G (1993) Adhesion to smooth surfaces by insects – a review. Int J Adhesion 13:3–7

Wick SM (1993) Immunolabeling of antigens in plant cells. Methods Cell Biol 37:172–199

Wyatt SE, Carpita N (1993) The plant cytoskeleton-exocellular matrix continuum. Trends Cell Biol 3:413–417

Yamamoto E, Bokelman G, Lewis NG (1989) Phenylpropanoid metabolism in cell walls, an overview. In: Lewis NG, Paice MG (eds) Plant cell wall polymers, biogenesis and degradation, Am Chem Soc Symp Ser 399:68–88

Young DH, Kauss H (1984) Adhesion of *Colletotrichum lindemuthianum* spores to *Phaseolus vulgaris* hypocotyls and to polystyrene. Appl Environ Microbiol 47:616–619

Zhu JK, Damsz B, Kononowicz AK, Bressar RA, Hasegawa PM (1994) A higher plant extracellular vitronectin-like adhesion protein is related to the translational elongation factor-1α. Plant Cell 6:394–404

Biochemical, Immunological and Molecular Analyses of Extensin

ELIZABETH E. HOOD

1 Introduction

The cell walls of higher plants contain a variety of glycoproteins including enzymes and structural proteins. One of the structural proteins is extensin, an hydroxyproline-rich glycoprotein. It is found in dicotyledonous and monocotyledonous angiosperms, as well as some gymnosperms (for review see: Cassab and Varner 1988; Wycoff et al. 1991; Showalter 1993). These proteins are ionically bound to the wall soon after secretion, but many become covalently cross-linked after short duration in muro (Cooper and Varner 1984; Fry 1987). Each plant and tissue type in which extensin is found has unique characteristics. However, common methods can be applied to study extensin from most tissues, and include biochemical characterization of the protein, use of antibodies for in vivo and in vitro detection, and molecular tools to study their genes. In this review, several of these methods are discussed.

2 Biochemical Characterization of Extensin

2.1 Preparation of Walls

Cell wall preparations that are free of other cellular materials are essential to draw conclusions concerning the proteins extracted from them. It is important that all cells are broken, a condition that can be monitored by viewing the homogenate under the light microscope. Tissues are ground in liquid nitrogen with a mortar and pestle, homogenized in a glass homogenizer on ice, or passed through a French pressure cell (2000 psi, 2–3 times) appropriate for suspension culture cells. Because walls are pelleted at each step of purification, it is important to centrifuge the preparation at low speed (~1000 × *g*) particularly during early stages to avoid pelleting organelles.

One wall preparation method (Atwell and Rees 1986) utilizes an extraction buffer containing 40 mM glycylglycine (pH 7.5), 1% (v/v) Triton-X 100, and 0.5 M KCl. For each 100 mg fresh weight of tissue ground, 1 ml of extraction buffer is used. After thorough mixing, walls are pelleted at 1000 × *g* for 10 min and resuspended in the same volume of extraction buffer before recentrifugation. This

Modern Methods of Plant Analysis, Vol. 17
Plant Cell Wall Analysis
Edited by H.-F. Linskens and J.F. Jackson

second pellet is resuspended in a volume of water equivalent to the initial fresh weight (e.g. 100 µl per 100 mg) and spun 10 min at a speed equivalent to approximately 15 000 × *g*, such as in a microfuge. This high-speed pellet can be resuspended and extracted immediately or frozen at −20 °C. For extraction of salt soluble proteins including extensin, the pellet is resuspended in a volume equal to the initial fresh weight of 0.2 M $CaCl_2$, 4 mM sodium metabisulfite (NaS_2O_5) for 2–12 h, spun, the supernatant collected and desalted as described below. For sodium dodecyl sulfate (SDS) extraction, the pellet is resuspended in a volume equivalent to the initial fresh weight of 1X Laemmli (1970) sample buffer, boiled 5 min, centrifuged 10 min, and the supernatant viewed on an SDS polyacrylamide gel.

A second method for cell wall preparation is described by Cassab et al. (1985). Tissues are ground as above but suspended in a buffer of 0.1% potassium acetate, 4 mM NaS_2O_5, and 1% insoluble polyvinyl pyrrolidone (PVP), an inhibitor of oxdiases that prevents tissue discoloration during wall preparation. The PVP is optional for tissues in which oxidases are not prevalent. Walls are pelleted at low speed (1000 × *g*, 10 min) and resuspended in five times the fresh weight (w/v) of 0.5% (v/v) Nonidet P-40, 2 mM NaS_2O_5, to disrupt membranes. After spinning, the pellet is washed 3–10 times in 2 mM NaS_2O_5 until supernatant is clear, at which point the cytoplasm is effectively removed. These walls are ready for salt extraction as described above. Higher yields are obtained if a short salt wash of the pellet is combined with the first salt extraction.

A third method for extracting extensin from walls does not require wall isolation. Tissue culture cells are used directly for rapid elution of ionically bound proteins after suspension in a 100 mM $AgCl_3$ salt solution followed by rapid filtration (Kieliszewski et al. 1990). Also, cultured cells can be packed into columns and wall proteins directly eluted with a NaCl gradient (Smith et al. 1984).

Extensin is a stable protein and fresh or frozen (at −20 or −80 °C) tissue may be used for its preparation from walls. If one begins the preparation with tissues rich in hydroxyproline or extensin such as aerated carrot root slices (Chrispeels et al. 1974), soybean seed coats (Cassab et al. 1985) or maize silk and pericarp (E.E. Hood et al. 1991; K.R. Hood et al. 1991), the amount of extensin relative to other proteins is increased and simplifies the purification.

2.2 Protein Purification

The $CaCl_2$ in the protein extracts can be removed via dialysis (Cassab et al. 1985; Kieliszewski and Lamport 1987) or a centrifuge column (K.R. Hood et al. 1991). For the column, the combined salt supernatants are concentrated by lyophilization and are resuspended in a small volume (usually 200 µl) of water. A spun column is prepared in a 3 ml syringe barrel plugged with glass wool and filled to capacity with Bio-gel P6DG (Bio-Rad). In our experience, this acrylamide resin does not bind the extensin, a problem encountered with carbohydrate beads or dialysis tubing, and we recover nearly equivalent amounts of hydroxyproline as applied to the column. The column is placed in a centrifuge tube such that the end of the syringe barrel is suspended above the bottom of the tube and the column is pre-spun in a swinging bucket rotor at 1000–2000 × *g* for 3 min. The sample is loaded onto the dried column and spun at 1000–2000 × *g* for 4–5 min, the sample being collected in a microfuge tube placed below the syringe barrel prior to spinning. Extensin can then be purified from the desalted protein extract.

Several laboratories have concentrated on the biochemical characterization of extensin and have developed purification methods for the protein. Extensins generally are soluble in 10% (w/v) trichloroacetic acid (TCA), whereas many other proteins, particularly cytoplasmic proteins, are not. Therefore, treatment of the salt-soluble extract with 10% TCA provides a partial purification. Kieliszewski and Lamport (1987) purified maize extensin via column chromatography first by phosphocellulose (Cellex-P, Bio-Rad) chromatography then by Biorex-70 (Bio-Rad) chromatography and lastly on Superose-6 (Pharmacia) beads. Fractions rich in hydroxyproline were collected at each step of purification. Similar methods were used to purify extensin from sugar beet (Li et al. 1990). Stafstrom and Staehelin (1987) purified carrot extensin by CM-cellulose chromatography and extensin fractions were eluted with a NaCl gradient. One impure extensin-containing fraction was further fractionated on Sephacryl S-400 in 150 mM Tris-HCl (pH 8). Hood et al. (1988) purified maize extensin by taking advantage of the basic nature of the protein and used chromatofocusing column chromatography, eluting with a pH gradient. Several charge variants of the protein of similar molecular weight were isolated (E.E. Hood, unpubl. data) though the most abundant was the most basic isoform which was further characterized. All investigators assessed protein purity by silver staining SDS-polyacrylamide gels (Laemmli 1970).

Preparative native gel electrophoresis can be performed to obtain large quantities of pure protein after excision and elution of the band. This method was used by Fritz et al. (1991) for monoclonal antibody preparation. The bicinchoninic acid protein assay (Smith et al. 1985; Pierce Chemical Co.) is an accurate way of determining the protein concentration of these hydroxyproline-rich proteins. In most cases, hydroxyproline content of hydrolyzed samples is determined by the colorimetric method of Drozdz et al. (1976).

2.3 Biochemical Characterization

Amino acid composition can assist the investigator in determining whether the protein being studied is likely to be an extensin before proceeding to the next more difficult level of investigation. Extensins from dicotyledons are rich in hydroxyproline, valine, tyrosine, serine, histidine and lysine, whereas those from monocotyledons, particularly the grasses, are rich in hydroxyproline, threonine, proline, serine and lysine (Showalter 1993). Amino acid composition is determined on acid-hydrolyzed protein fractions (Kieliszewski and Lamport 1987; Hood et al. 1988) and can be used to monitor increases in specific amino acids during successive purification steps. Proteins such as extensins with biased amino acid compositions lend themselves to labeling in vivo with one of those amino acids, proline, for example (Kleis-San Francisco and Tierney 1990). A modification of this method which involves treating tissues with various labeled amino acids can be utilized to discover new proteins with biased amino acid compositions.

Amino acid sequences of proteins are valuable for learning the relatedness of individual extensins from the same plant and for comparing them between plants. Amino-terminal sequence can be obtained from nanogram quantities of purified protein in solution or blotted onto membranes from gels (Matsudaira 1987). Whole protein sequence can be obtained from overlapping cleavage products separated on HPLC and subjected to Edman degradation (e.g., see Kïeliszewski et al. 1990).

Extensins are glycoproteins containing arabinoses and galactoses O-linked to hydroxyproline and serine, respectively. Two types of information about glycoproteins are better obtained on deglycosylated protein products than on holoproteins – identification of amino acids to which sugars are attached, and molecular weight determination of the protein backbone. Therefore, the anhydrous hydrogen fluoride method for removing O-linked sugars (Mort and Lamport 1977) is often employed in extensin studies. This method leaves the peptide linkages intact, allowing characterization of the deglycosylated products. The sugars removed can also be identified, an important piece of biochemical information.

Extensins that are newly secreted into the cell wall are ionically bound and soluble in salt solutions. However, after some duration within the wall, extensins beocme resistant to salt extraction. It appears that covalent cross-links between and among extensin molecules are being formed (Cooper and Varner 1984), most likely isodityrosine (IDT) cross-links between adjacent tyrosine residues (McNeil et al. 1984; Fry 1986). This conclusion was reached because insoluble extensins can be solubilized by treating walls with acidified $NaClO_2$ which cleaves IDT linkages, but not peptide bonds (Fry 1986). Acid hydrolysis of cell walls allows isolation of IDT, a phenolic compound, that is identified through thin layer chromatography (Fry 1984). Some evidence leads to the suggestion that a peroxidase specific for extensin cross-linking is present in walls of some plants (Everdeen et al. 1988). Thus, an exciting area for further study in extensin biology is to characterize the extensin-specific peroxidase.

3 Immunological Detection of Extensin

Polyclonal and monoclonal antibodies can be generated against the native (Stafstrom and Staehelin 1988; Fritz et al. 1991) and deglycosylated extensins (Kieliszewski and Lamport 1987; Stafstrom and Staehelin 1988) and against synthetic peptides generated from gene sequence (Steifel et al. 1988). Occasionally, anti-extensin antibodies cross-react with a number of other glycoproteins because the extensin molecule is coated with carbohydrate (Showalter 1993) and antibodies directed against those sugars will recognize the same sugars on other proteins. Nevertheless, as long as proper controls are performed, antibodies are a powerful and specific tool to learn where and when the protein is expressed.

3.1 In Vivo Localization of Extensin

Antibody labeling in conjunction with light and electron microscopy has been employed to elucidate the location of extensin in plant tissues. Care should be taken to embed tissues in partially hydrophilic resins such as JB-4, LR White and Spurr, which maximize antibody reactivity. In thick sections prepared for light microscopy, primary antibodies are localized using a secondary antibody conjugated to a detectable compound such as alkaline phosphatase, horseradish peroxidase, gold particles enhanced with silver, or a fluorescent moiety. A typical protocol for labeling tissues with alkaline-phosphatase conjugated secondary antibodies is outlined in Fritz et al. (1991), for silver-enhanced gold labeling in Cassab and Varner (1987), and for fluorescence detection in Leach et al. (1982) and Laxer et al. (1987). In ultrathin sections prepared for electron microscopy, the secondary antibody should be conjugated to an electron dense particle, most often colloidal gold (Bonfante-Fasolo et al. 1991). Smaller gold particles may be enhanced with silver, a technique that prevents steric hindrance from larger gold particles during binding, but allows highly sensitive detection of the small particles. Typical protocols for detecting primary antibody location with the electron microscope are described by Stafstrom and Staehelin (1988) and O'Connell et al. (1990). An excellent review of localization methods utilized to study extensin structure are described by Swords and Staehelin (1989) in Volume 10 of this series.

3.2 Western Blots and Tissue Prints

Unpurified salt-extracted proteins from tissues can be separated on either native (e.g., Thomas and Hodes 1981) or denaturing (Laemmli 1970) polyacrylamide gels and electroblotted onto membranes such as Immobilon (Millipore) (K.R. Hood et al. 1991). A typical blotting buffer contains 48 mM Tris, pH 9.2, 39 mM glycine, 1.3 mM SDS and 20% methanol (Gelman Sciences Application Protocols). Western blots are highly sensitive and can be used to detect small amounts of extensin in mixes of protein extracted from cell walls. These quantitative blots are used to compare extensin in equal masses of total protein from various tissues. In addition, they are used to assess antigenic relatedness and molecular weights of proteins in different extracts (Fig. 1).

Tissue printing is a quick assay to determine the tissue location of extensin (Cassab 1992). To perform a tissue print, freshly cut plant tissue is held in firm contact with nitrocellulose for 5 to 30 s. The filter paper should be padded with a stack of paper towels so that hard areas of the printed tissue actually make depressions in the filter. Filter papers soaked in 0.2 M sodium chloride or calcium chlo-

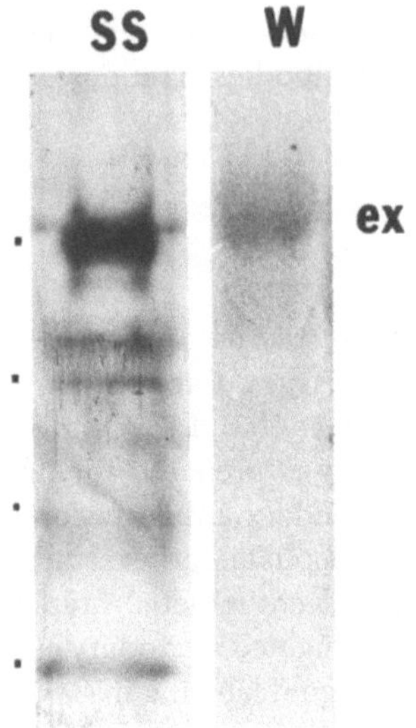

Fig. 1. Maize silk cell wall proteins size separated on an SDS polyacrylamide gel are stained with silver nitrate (*SS*). Of the multiple protein bands, only one at approximately 66 kDa is detected with anti-extensin monoclonal antibodies on a western blot (*W*). *Dots to the left* and from *top to bottom* represent molecular weights of marker proteins: bovine serum albumin, 66 kDa; ovalbumin, 45 kDa, trypsinogen, 36 kDa, and cytochrome C at 22 kDa; *ex* extensin

ride provide an extracting surface for the proteins, which bind to the filter when they come in contact with it. Tissues from different parts of the plant or from different developmental stages can be compared on a single filter containing multiple prints. An example of this method (Fig. 2) shows prints of three varieties of maize kernels from 10 and 20 days after pollination in which extensin is detected primarily in the pericarp. In this example, the extensin antibody was detected with an alkaline phosphatase-conjugated secondary antibody and nitroblue tetrazolium and bromo-chloro-indolyl phosphate as substrates (described in Fritz et al. 1991). The major advantage of tissue printing is the ability to assay whole plant tissues and organs for accumulation of extensin without extracting the protein. The disadvantage is that only the soluble extensins are detectable.

4 Molecular Characterization of Extensin

Extensin is important in normal development, wounding, and defense against pathogens. Studies of extensin expression during these diverse processes have been simplified by measuring the accumulation of its mRNA rather than quantifying the protein. For an insoluble protein such as extensin, this is sometimes the most accurate measure. Molecular probes can be used on tissue prints to assess the cell types where extensin mRNA is abundant (McClure and Guilfoyle 1989; Ye and Varner 1991)). Promoter fusions with reporter genes provide specific information on which cell types are expressing extensin (Wycoff et al. 1990). Several examples of molecular studies on extensins are included in a recent review (Showalter and Rumeau 1990).

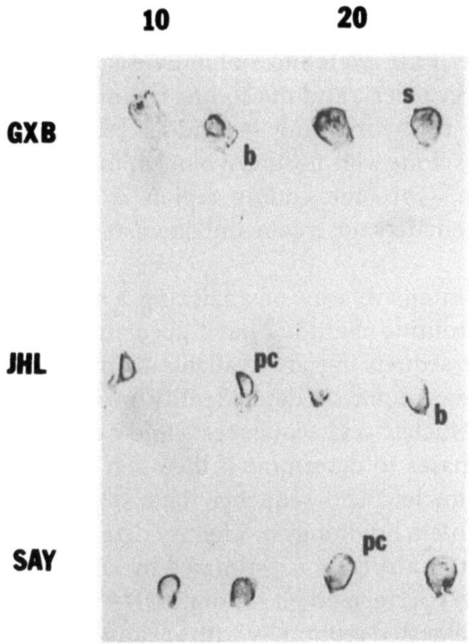

Fig. 2. Tissue print analysis of maize kernels from three varieties of maize. Whole kernels at 10 and 20 days after pollination were cut vertically with a razor blade and pressed onto nitrocellulose for 30 s; *b* base of the ketnel where it attaches to the cob, *s* silk scar; *pc* pericarp. The nitrocellulose was blocked with 3% BSA in Tris-buffered saline Tween (Fritz et al. 1991), treated with anti-extensin primary antibody (1:10 000 dilution) and goat anti-mouse IgG alkaline phosphatase conjugate (1:10 000; Sigma Chemical Co.). Alkaline phosphatase was detected with nitroblue tetrazolium and bromo-chloro-indolyl phosphate. *G×B* Golden cross Bantam sweet corn; *JHL* Japanese hull-less popcorn; *SAY* South American yellow popcorn

4.1 Isolation of Genes Encoding Extensin

The production of a cDNA library is standard and is described in Sambrook et al. (1989). Moreover, several kits for library construction are commercially available. Methods to screen a cDNA library include using heterologous probes, oligonucleotides based on protein sequence, and antibodies for which an expression library is needed. Tissues that have abundant extensin mRNA are logical choices for preparing a cDNA library from which extensin clones are to be isolated. Alternatively, a genome library can be prepared to isolate full length clones for extensin genes. Again, numerous kits and methods are available for this activity. In addition to the probes suggested above, a cDNA previously isolated is an excellent probe to use for screening a genome library. Antibodies cannot be used to screen genome libraries because eukaryotic genes are not expressed in *E. coli*.

Because extensin is often represented by a multigene family, more than one distinct clone for extensin is likely present in a genome library. When more than one gene is expressed in a tissue, this is true also for cDNA libraries. Multiple clones can be divided into classes based on digestion patterns generated by restriction endonucleases. Cross-hybridization among clones and expression studies using gene-specific probes can further determine relatedness of individual clones. Genomic clones generally contain quite large inserts and the coding region should be identified. This can be accomplished by hybridizing individual restriction endonuclease digestion fragments from the clone with northern blots of total RNA isolated from tissues expressing extensin. Once the coding region is located, sequencing of this region and regions immediately upstream and downstream will characterize the active parts of the gene.

The most informative though labor-intensive way of analyzing a clone is through nucleic acid sequencing. These routine methods have been published (Sambrook et al. 1989) and many kits for sequencing are available from any of several biotechnology products companies. Sequence data identify promoters, coding regions, and poly-A addition sites. Nucleic acid sequences on new extensin genes can be compared to others on databases to determine if the open reading frames and promoters are similar. From nucleic acid sequence data, the amino acid sequence can be predicted and thus protein homology or identity determined.

The number of extensin genes present in a species is estimated by DNA blot hybridization analysis (Sambrook et al. 1989) performed on genomic DNA isolated from the plant in question (Fig. 3). DNA is digested separately with various restriction endonucleases, size separated on agarose gels, and transferred to a membrane. To estimate the number of genes, additional gel lanes should contain the extensin gene at a concentration that reconstructs the hybridization intensity for one, five

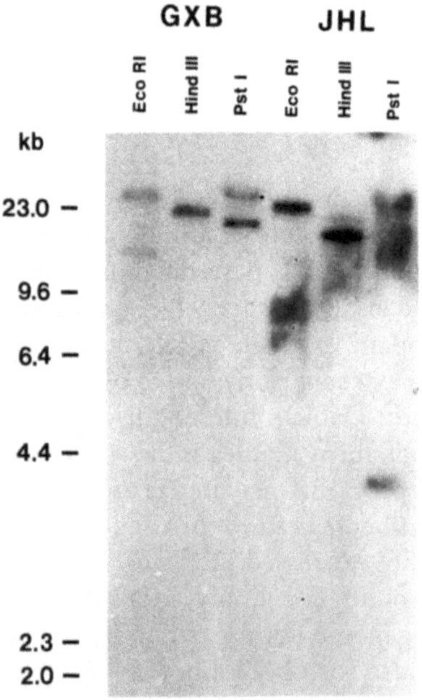

Fig. 3. DNA blot hybridization analysis of maize extensin genes. Ten µg of DNA was digested with the indicated restriction endonucleases and the fragments size separated on an agarose gel before transferring to a charged nylon membrane. The membrane was hybridized with a ^{32}P-labeled maize extensin gene (Steifel et al. 1988). *Numbers to the left* indicate sizes in kilobases of lambda DNA digested with *Hind*III. *G×B* Golden cross Bantam sweet corn; *JHL* Japanese hull-less popcorn. Restriction fragment length polymorphisms are apparent in extensin genes between these two varieties

and ten copies of the gene per microgram of plant DNA loaded. The membranes are hybridized with labeled extensin genes, and one counts not only the number of bands present, but compares the intensity of hybridization of those bands to the reconstruction lanes (e.g., see Hood et al. 1993). Of course, not all copies of genes in a family are likely to be expressed. However, the potential for expression can be predicted by analyzing the clone for promoter and terminator sequences and open reading frames. Alternatively, isolating cDNAs from tissues is the best evidence for expression of individual extensin genes.

4.2 Extensin Expression Studies

The accumulation of extensin mRNA can be assessed by several methods including northern blot analysis, dot blot analysis, tissue printing, and in vitro translation. Total RNA is isolated from plant cells by any of several published methods (e.g., Haffner et al. 1978). For northern blots, RNAs are size separated on denaturing agarose gels and transferred to a membrane, for example charged nylon. The filter is hybridized in standard solutions with ^{32}P-labeled probes most commonly generated by random primer labeling (Sambrook et al. 1989). Accumulation of extensin mRNA in various maize tissues was studied using northern and dot blot analyses (Hood et al. 1993) to learn where the greatest abundance of extensin mRNA is found. In vitro translation analyses were utilized to study the size of maize extensin protein products prior to glycosylation (Murphy and Hood 1993).

Studies of extensin expression have been done that document the genes' responses to external stimuli. These stimuli include: developmental signals, ehtylene, wounding, elicitors, gravity, and pathogen invasion (Showalter and Rumeau 1990). The accumulation of extensin mRNA in these cases provides evidence for extensin's role in the response of the plant to these stimuli. However, further and more detailed studies, such as those described in the next section, are necessary to define the actual role of extensin.

4.3 Extensin Promoter Fusions with Reporter Genes

An exciting new area in plant gene expression studies involves the translationally active fusion of a promoter with a reporter gene such as β-glucuronidase, a technique used to study extensin genes (e.g., Wycoff et al. 1990). Because this technique uses a substrate, X-gluC, which turns blue in the presence of the enzyme, it is highly sensitive and can detect tissue-specific expression of a promoter at a level not attainable with mRNA analysis or in situ hybridization. Promoter fusions are particularly valuable to assess the expression of individual members of a multigene family for which gene-specific probes may not be available. Studies such as these will lead to a much more complete understanding of the temporal and tissue-specific regulation of extensin genes.

4.4 Generation of Extensin Mutants

Mutations in specific genes are the best evidence for the role of a protein in an organism. Mutations in extensin genes are no exception, and extensin mutants would lead to an understanding of their role in the plant cell wall. However, because extensin is part of a multigene family and because most plants have exceptionally large genomes, this experiment has been difficult. The small genome of *Arabidopsis thaliana* is amenable to generation of mutations, though extensin mutants would be difficult to identify because no known enzymatic function for the protein is available for which to screen. Therefore, one approach to generating mutations in extensin genes is to transform plants of interest with either antisense genes or overexpressing genes to assess the affects of lesser or greater amounts of the protein, respectively. This approach has been used effectively with polygalacturonase expression in tomato fruits (Picton et al. 1993) and will likely be applicable to extensin.

5 Summary

The cell wall is a vitally important part of plant form and function. Though the proteins are a minor part of the wall mass, they are essential to establishing the components, structure and cross-links within the wall. Extensins are one of the many structural proteins of plant walls and have been studied quite extensively in a few plants. However, until more data are collected on their biochemistry, localization and molecular biology, we will not fully understand their function. This review presented an array of methods available to study these interesting and important proteins.

Acknowledgments. I thank Jenifer M. Murphy, Robert C. Pendleton, and Sue E. Fritz for contributing data. I thank Gary Pedersen for help in assembling the figures.

References

Atwell BJ, Rees T (1986) Distribution of protein synthesized by seedlings of *Oryza sativa* grown in anoxia. J Plant Physiol 123:401–408

Bonfante-Fasolo P, Tamagnone L, Peretto R, Esquerre-Tugaye MT, Mazau D, Mosiniak M, Vian B (1991) Immunocytochemical location of hydroxyproline rich glycoproteins at the interface between a mycorrhizal fungus and its host plants. Protoplasma 165:127–138

Cassab GI (1992) Localization of cell wall proteins. In: Reid PD, Pont-Lezica RF, del Campillo E, Taylor R (eds) Tissue printing, tools for the study of anatomy, histochemistry, and gene expression. Academic Press, San Diego, p 188

Cassab GI, Varner JE (1987) Immunocytolocalization of extensin in developing soybean seed coats by immunogold-silver staining and by tissue printing on nitrocellulose paper. J Cell Biol 105:2581–2588

Cassab GI, Varner JE (1988) Cell wall proteins. Annu Rev Plant Physiol Plant Mol Biol 39:321–353
Cassab GI, Nieto-Sotelo J, Cooper JB, Van Holst G-J, Varner JE (1985) A developmentally regulated hydroxyproline-rich glycoprotein from the cell walls of soybean seed coats. Plant Physiol 77:532–535
Chrispeels MJ, Sadava D, Cho YP (1974) Enhancement of extensin biosynthesis in aging discs of carrot storage tissue. J Exp Bot 25:1157–1166
Cooper J, Varner JE (1984) Cross-linking of soluble extensin in isolated cell walls. Plant Physiol 76:414–417
Drozdz M, Kucharaz E, Szyja J (1976) A colorimetric micromethod for determination of hydroxyproline in blood serum. Z Med Labortechnik 17:163–171
Everdeen DS, Kiefer S, Willard JJ, Muldoon EP, Dey PM, Li X-B, Lamport DTA (1988) Enzymic cross-linkage of monomeric extensin precursors in vitro. Plant Physiol 87:616–621
Fritz SE, Hood KR, Hood EE (1991) Localization of soluble and insoluble fractions of hydroxyproline-rich glycoproteins during maize kernel development. J Cell Sci 98:545–550
Fry SC (1984) Isodityrosine – its detection, estimation and chemical synthesis. Methods Enzymol 107:388–397
Fry SC (1986) Cross-linking of matrix polymers in the growing cell walls of angiosperms. Annu Rev Plant Physiol 37:165–186
Fry SC (1987) Formation of isodityrosine by peroxidase isozymes. J Exp Bot 38:853–862
Haffner MH, Chin MB, Lane BG (1978) Wheat embryo ribonucleates. XII. Formal characterization of terminal and penultimate nucleoside residues at the 5′-ends of "capped" RNA from imbibing wheat embryos. Can J Biochem 56:729–733
Hood EE, Shen QX, Varner JE (1988) A developmentally regulated hydroxyproline-rich glycoprotein in maize pericarp cell walls. Plant Physiol 87:138–142
Hood EE, Hood KR, Fritz SE (1991) Hydroxyproline-rich glycoproteins in cell walls of pericarp from maize. Plant Sci 79:13–22
Hood EE, Murphy JM, Pendleton RC (1993) Molecular characterization of maize extensin expression. Plant Mol Biol 21:885–893
Hood KR, Baasiri RA, Fritz SE, Hood EE (1991) Biochemical and tissue print analyses of hydroxyproline-rich glycoproteins in cell walls of sporophytic maize tissues. Plant Physiol 96:1214–1219
Kieliszewski MJ, Lamport DTA (1987) Purification and partial characterization of a hydroxyproline-rich glycoprotein in a graminaceous monocot, *Zea mays*. Plant Physiol 92:316–326
Kieliszewski MJ, Leykam JF, Lamport DTA (1990) Structuare of the threonine-rich extensin from *Zea mays*. Plant Physiol 92:316–326
Kleis-San Franscisco SM, Tierney ML (1990) Isolation and characterization of a proline-rich cell wall protein from soybean seedlings. Plant Physiol 94:1897–1902
Laemmli UK (1970) Cleavage of structural proteins during assembly of the head of bacteriophage T4. Nature 227:680–685
Laxer M, Healey M, Youssef NN (1987) Production of monoclonal antibodies specific for *Eimeria tenella* microgametocytes. J Parasitol 73:611–616
Leach JE, Cantrell MA, Sequeira L (1982) A hydroxyproline-rich bacterial agglutinin from potato: its localization by immunofluorescence. Physiol Plant Pathol 21:319–325
Li X-B, Kieliszewski M, Lamport DTA (1990) A chenopod extensin lacks repetitive tetrahydroxyproline blocks. Plant Physiol 92:327–333
Matsudaira P (1987) Sequence from picomole quantities of proteins electroblotted onto polyvinylidene difluoride membranes. J Biol Chem 262:10035–10038
McClure BA, Guilfoyle TJ (1989) Tissue print hybridization. A simple technique for detecting organ- and tissue-specific gene expression. Plant Mol Biol 12:517–524
McNeil M, Darvill AG, Fry SC, Albersheim P (1984) Structure and function of the primary cell walls of plants. Annu Rev Biochem 53:625–663
Mort AJ, Lamport DTA (1977) Anhydrous hydrogen fluoride deglycosylated glycoproteins. Anal Biochem 82:289–309
Murphy JM, Hood EE (1993) Molecular basis for extensin size heterogeneity in two maize varieties. Plant Mol Biol 21:885–893

O'Connell RJ, Brown IR, Mansfield JW, Bailey JA, Mazau D, Rumeau D, Esquerre-Tugaye M-T (1990) Immunocytochemical localization of hydroxyproline-rich glycoproteins accumulating in melon and bean at sites of resistance to bacteria and fungi. Mol Plant-Microbe Interact 3:33–40

Picton S, Barton SL, Bouzayen M, Hamilton AJ, Grierson D (1993) Altered fruit ripening and leaf senescence in tomatoes expressing an antisense ethylene-forming enzyme transgene. Plant J 3:469–481

Sambrook J, Maniatis T, Fritsch EF (1989) Molecular cloning: a laboratory manual, 2nd edn, vol 3. Cold Spring Harbor Laboratory, Cold Spring Harbor

Showalter AM (1993) Structure and function of plant cell wall proteins. Plant Cell 5:9–23

Showalter AM, Rumeau D (1990) Molecular biology of plant cell wall hydroxyproline-rich glycoproteins. In: Adair WS, Mecham RP (eds) Organization and assembly of plant and animal extracellular matrix. Academic Press, San Diego, pp 247–281

Smith JJ, Muldoon EP, Lamport DTA (1984) Isolation of extensin precursors by direct elution of intact tomato cell suspension cultures. Phytochemistry 23:1233–1239

Smith PK, Krohn RI, Hermanson GT, Mallia AK, Gartner FH, Provenzano MD, Fujimoto EK, Goeke NM, Olson BJ, Klenk DC (1985) Measurement of protein using bicinchoninic acid. Anal Biochem 150:76–85

Stafstrom JP, Staehelin LA (1987) A second extensin-like hydroxyproline-rich glycoprotein from carrot cell walls. Plant Physiol 84:820–825

Stafstrom JP, Staehelin LA (1988) Antibody localization of extensin in cell walls of carrot storage roots. Planta 174:321–322

Steifel V, Perez-Grau L, Albericio F, Giralt E, Ruiz-Avila L, Ludevid MD, Puigdomenech P (1988) Molecular cloning of cDNAs encoding a putative cell wall protein from *Zea mays* and immunological identification of related polypeptides. Plant Mol Biol 11:483–493

Swords KMM, Staehelin LA (1989) Analysis of extensin structure in plant cell walls. In: Linskens HF, Jackson JF (eds) Modern methods of plant analysis, vol 10. Springer, Berlin Heidelberg New York, pp 219–233

Thomas JM, Hodes ME (1981) A new discontinuous buffer system for electrophoresis of cationic proteins at near-neutral pH. Anal Biochem 118:194–196

Wycoff KL, Powell PA, Dixon RA, Lamb CJ (1990) Tissue-specific and would responsive expression of a hydroxyproline-rich glycoprotein gene promoter in transgenic tobacco. J Cell Biochem (Suppl) 14E:344

Wycoff KL, Dixon RA, Lamb CJ (1991) Hydroxyproline-rich glycoproteins in plant-microbe interactions and development. In: Verma DPS (ed) Molecular signals in plant microbe communications. CRC Review, CRC Press, Boca Raton, p 521

Ye ZH, Varner JE (1991) Tissue-specific expression of cell wall proteins in developing soybean tissues. Plant Cell 3:23–37

Analysis of Pectin Structure by HPAEC-PAD

ARLAND T. HOTCHKISS, JR., KARIMAH EL-BAHTIMY, and
MARSHALL L. FISHMAN

1 Introduction

Pectins are complex, branched, acidic polysaccharides that are located throughout the plant primary cell wall matrix. Pectin structure in cell walls has been described as block-like (Jarvis 1984) in which smooth and hairy regions exist (de Vries et al. 1982). The smooth pectin regions consist of homogalacturonan (McNeil et al. 1984). Variation in the degree of methyl-esterification exists within pectin, with calcium mediating the cooperative cross-linking of pectin chains containing seven adjacent nonesterified galacturonic acid residues (Powell et al. 1982). When pectin is treated with endo-polygalacturonase (endo-PG[1]), the homogalacturonan degrades to oligogalacturonic acids leaving the hairy, branched pectic regions unhydrolyzed (reviewed by O'Neill et al. 1990). The hairy regions of pectin consist of rhamnogalacturonan I (RG-I), rhamnogalacturonan II (RG-II), xylogalacturonan, and RG with arabinan side chains that can be separated based on size following endo-polygalacturonase (O'Neill et al. 1990) or rhamnogalacturonase (Voragen et al. 1993) hydrolysis. RG-I consists of alternating rhamnose and galacturonic acid residues in the backbone. Arabinose- and galactose-containing side chains are attached to the rhamnose residues. RG-II, from the cell walls of suspension-cultured sycamore cells, consists of a not fully characterized structure that contains unusual sugars such as apiose, 3-deoxy-D-manno-octulosonic acid (KDO), aceric acid, methyl-fucose, and methyl-xylose. Xylogalacturonan, present in the modified hairy regions (MHR, residue remaining after enzymatic liquefaction of pectin) of apple pectin, consists of a galacturonan backbone with each galacturonic acid residue substituted at the 3-position with a single β-linked xylose

---

Mention of brand or firm names does not constitute an endorsement by the US Department of Agriculture over others of a similar nature not mentioned.

[1] Abbreviations: ASP, alkaline soluble pectin; CDTA, 1,2-cyclohexanediaminetetraacetic acid; CSP, chelator-soluble pectin; DEAE, diethylaminoethyl; DP, degree of polymerization; endo-PG, endo-polygalacturonase; FAB-MS, fast atom bombardment-mass spectrometry; GPC, gel-permeation chromatography; HPAEC-PAD, high-performance anion-exchange chromatography and pulsed amperometric detection; HPLC, high-performance liquid chromatography; HPSEC, high-performance size-exclusion chromatography; MHR, modified hairy regions; MHR-S, saponified modified hairy regions; MS, mass spectrometry; NMR, nuclear magnetic resonance; PAD, pulsed amperometric detection; PME, pectin methyl-esterase; QAE, diethyl[2-hydroxypropyl]aminoethyl; RG, rhamnogalacturonan; UV, ultraviolet; WAF, weeks after flowering.

Modern Methods of Plant Analysis, Vol. 17
Plant Cell Wall Analysis
Edited by H.-F. Linskens and J.F. Jackson

residue (Voragen et al. 1993). After approximately ten galacturonic acid residues the xylogalacturonan backbone is interrupted by a rhamnose residue to which arabinose- and galactose-rich side chains may be attached. RG with arabinan side chains was also characterized in apple MHR as smaller in molecular weight compared to xylogalacturonan but larger than RG-I (Voragen et al. 1993). Curve-fitting of high-performance size-exclusion chromatograms indicated the presence of five macromolecular components for pectins from various sources (Fishman et al. 1989b, 1992). Electron microscopy of shadowed pectic components demonstrated a multimodal distribution of chain lengths (Fishman et al. 1992; McCann et al. 1992). Macromolecular pectic components are thought to represent different levels of noncovalent aggregation of these acidic polysaccharides. In this chapter, we demonstrate that enzymatic degradation of pectin subunits followed by high-performance anion-exchange chromatography and pulsed amperometric detection (HPAEC-PAD) analysis yields valuable information concerning the internal structure of these pectic subunits.

2 Pectin HPLC Separations

2.1 GPC

Gel-permeation (GPC) or size-exclusion chromatography has long been used to separate pectins based on size. Determination of pectin molecular weight is difficult with this type of chromatography due to aggregation. However, high-performance size exclusion chromatography (HPSEC) has been used to measure pectin molecular weight and size (Barth 1980; Fishman et al. 1984, 1987; Deckers et al. 1986), as well as to monitor pectin depolymerization by pectate lyase (Neidhart et al. 1994). These columns are calibrated with either pullulan, dextran, or pectin standards. HPSEC columns require long run times and possess relatively low oligosaccharide resolution compared to other forms of high-performance liquid chromatography (HPLC). Following the first reports of pectic oligosaccharide resolution with gel filtration chromatography (Rexova-Benkova 1970; Thibault 1980), Bio-Gel P-4 has recently been used to purify oligogalacturonic acids up to a degree of polymerization (DP) of 6 (MacDougall et al. 1992) and to monitor pectin molecular weight changes during fruit repening in avocado and tomato (Huber and O'Donoghue 1993). While the resolution of the P-4 column is better than other gel-filtration or size-exclusion columns, the run times are still rather long (8–11 h).

2.2 Ion-Exchange and Ion-Pair RP

For over 40 years, ion-exchange chromatography has been used to separate pectins and their degradation products (Derungs and Deuel 1954). Separation of pectin from neutral polysaccharides has typically involved ambient pressure ion-exchange chromatography on DEAE Sepharose or Cellulose (de Vries et al. 1981; Rouau and Thibault 1984), and QAE-Sephadex (Nothnagel et al. 1983). High-performance ion-exchange chromatography has been used to separate oligogalacturonic acids (Voragen et al. 1982) with a DP of up to 25 (Maness and Mort 1989), to isolate oligogalacturonic acids up to DP 7 in gram quantities (Hotchkiss et al. 1991), to determine the degree of esterification of pectin and monitor deesterification with pectin methyl esterase (Schols et al. 1989), and to monitor enzymatic depolymerization of polygalacturonic acid (Endress et al. 1991). While the run times are reasonable (<1 h), this method requires the use of a UV-absorbance detector which produces a significant baseline shift during the course of the buffer gradient. In order to resolve oligogalacturonic acids above approximately DP 10, it is necessary to first form 2-amino-pyridine derivatives (Maness and Mort 1989). Ion-pair reversed-phase chromatography has also been used to separate oligogalacturonic acids (Heyraud and Rochas 1982; Voragen et al. 1982) and monitor depolymerization of polygalacturonic acid by pectate lyase (Preston and Rice 1991). However, oligogalacturonic acids greater than DP 11 could not be resolved by this method. Additionally, the column effluent, monitored by UV absorbance, was subject to baseline shifts.

2.3 HPAEC-PAD

HPAEC-PAD is the most selective HPLC method available for the separation of monosaccharides, oligosaccharide positional isomers, and homopolymer series of oligosaccharides (Hardy and Townsend 1988; Koizumi et al. 1989; Lee 1990). This method allows for the separation of underivatized oligosaccharides up to DP 80 (Koizumi et al. 1989). The detector is as sensitive as the highest sensitivity refractive index HPLC detectors, while almost insensitive to changes in mobile phase buffer concentration. Hotchkiss and Hicks (1989, 1990) first applied HPAEC-PAD to the separation of pectic oligosaccharides. Subsequently, numerous publications have appeared demonstrating the value of HPAEC-PAD in structural analysis of pectins.

A series of oligogalacturonic acids terminated by a 4,5-unsaturated function at their nonreducing ends was observed to elute later from the CarboPac PA1 column used for HPAEC-PAD than the corresponding like-DP saturated counterparts (Hotchkiss et al. 1990; Hotchkiss and Hicks 1993; Lieker et al. 1993). This allowed for HPAEC-PAD analysis of pectate lyase (highly purified and "recombinant") action patterns demonstrating that both unsaturated and saturated oligogalacturonic acids were generated from polygalacturonic acid and plant cell walls during a depolymerization time course (Hotchkiss and Hicks 1993; Hotchkiss et al. 1995b). Oligogalacturonic acids were retained longer on the CarboPac PA1 column than the same DP oligosaccharide consisting of galacturonic acid residues terminated at the reducing end by a single galactosyl residue (Mort et al. 1993). In order to analyze the pattern of pectin methyl-esterification, the latter oligosaccharides were formed by reduction of methyl-esterified galacturonic acid residues followed by HF solvolysis. HPAEC-PAD was also used to identify oligogalacturonic acids terminated at their reducing ends with galactaric acid isolated from commercial polygalacturonic acid (Pressey 1991). Uronic acid oxidase was found in squash and other plants (Pressey 1991), therefore, oxidized oligogalacturonic acids are likely cell wall components that are not generated by microbial attack. Since these oxidized oligogalacturonic acids stimulated auxin hormone oxidation (Pressey 1991), they may be plant signal compounds that have greater auxin antagonist activity than unoxidized oligogalacturonic acids (Fry et al. 1993).

HPAEC-PAD was used for the semipreparative (9 × 250 mm CarboPac PA1 column) isolation of oligogalacturonic acids between DP 10 and DP 15 (Spiro et al. 1993). The structure of DP 13 oligogalacturonic acid, a phytoalexin elicitor, was confirmed by ^{1}H-NMR and FAB-MS. Lo et al. (1994) used this semipreparative HPAEC-PAD method and reported ^{1}H- and ^{13}C-NMR chemical shift values and ^{1}H-^{1}H coupling constants for isolated oligogalacturonic acids up to DP 7 as well as a FAB-MS $[M-H]^+$ molecular ion for DP 14. Melotto et al. (1994) used a preparative-scale CarboPac PA1 column (23 × 280 mm) to isolate tomato (breaker stage of ripeness) rhamnogalacturonan oligosaccharides that elicited ethylene production in mature green tomato pericarp discs.

Schols et al. (1994b) used semipreparative HPAEC-PAD to isolate rhamnogalacturonan oligosaccharides up to DP 9 from rhamnogalacturonase-degraded, saponified, modified hairy regions (MHR-S) of apple pectin. ^{1}H-NMR spectroscopy of the isolated oligosaccharides confirmed the structure of the RG-I disaccharide repeating unit, α-(1 → 2)-L-Rha*p*-α-(1 → 4)-D-Gal*p*A (with approximately half of the rhamnose residues substituted at the 4-position with β-Gal*p*; Lau et al. 1985). This structure was reported earlier (Colquhoun et al. 1990) following NMR analysis of a mixture of rhamnogalacturonase-generated oligosaccharides in a low molecular weight fraction collected from a Sephadex G50 GPC column (Schols et al. 1990a). Molecular ions ($[M + Na]^+$) of these rhamnogalacturonan oligosaccharides were reported following HPAEC-thermospray MS (Schols et al. 1994a). Cell wall RG-I susceptible to the fungal rhamnogalacturonase (first purified by Schols et al. 1990a) appears to have wide taxonomic source range. A similar profile of rhamnogalacturonan oligosaccharides was generated from dicot (apple, potato, carrot, pear) and monocot (leek and onion) MHR-S pectin following rhamnogalacturonase treatment (Schols et al. 1994b). A different rhamnogalacturonase (RGase B) was reported by Kofod et al. (1994) that was a lyase (RG-lyase) based on HPAEC-PAD, UV (235 nm) detection and ^{1}H-NMR of the rhamnogalacturonan oligosaccharides released (Mutter et al. 1996).

The structure of RG-I (O'Neill et al. 1990; Schols et al. 1990b) was further characterized when HPAEC-PAD was used to separate a series of arabinose-containing oligosaccharides up to DP 20 following treatment of sycamore suspension-culture RG-I with endo-(1 → 5)-α-L-arabinase (Lerouge et al. 1993). A combination of endo- and exo-(1 → 4)-β-D-galactanases was required to remove most of the galactose-containing side chain region from de-arabinosylated RG-I.

Galacturonic acid (Lerouge et al. 1993), glucuronic acid, and 4-*O*-methyl-glucosyluronic acid (An et al. 1994a) were observed to be substituents on the galactose-containing side chain of RG-I. The latter two substituents were components of side chain oligosaccharides isolated by semipreparative HPAEC-PAD following acid hydrolysis of sycamore RG-I. An et al. (1994b) used semipreparative HPAEC-PAD to isolate rhamnogalacturonan oligosaccharides (up to DP 14) generated from sycamore RG-I after rhamnogalacturonase activity was observed in a mixture of commercial cell wall-degrading enzymes. Mutter et al. (1994) purified a fungal α-L-rhamnosidase specific for apple rhamnogalacturonan fragments and characterized its mode of action by HPAEC-PAD. This enzyme removes the nonreducing terminal rhamnose residue from rhamnogalacturonan oligosaccharides. The HPAEC-PAD peak area of the series of rhamnogalacturonan oligosaccharides with nonreducing ends terminated by galacturonic acid was greater than the peak area of those oligosaccharides with rhamnose residues at their nonreducing ends when both rhamnogalacturonase and rhamnosidase activities were present during sycamore RG-I hydrolysis (An et al. 1994b).

Most of the above methods for HPAEC-PAD analysis of pectin structure have utilized a high pH mobile phase (100 mM NaOH) and a sodium acetate gradient. The limit of DP size resolution using these conditions is a DP 20 oligogalacturonic acid (O'Neill et al. 1990). Using a potassium oxalate (pH 6) gradient mobile phase and post-column addition of hydroxide, it is possible to resolve oligogalacturonic acids up to approximately DP 50 (Hotchkiss and Hicks 1990). Since oligogalacturonic acids up to DP 50 were not isolated and characterized, it was debatable whether or not these HPAEC-PAD peaks represented 50 covalently connected galacturonosyl residues without rhamnose interruption. Koizumi et al. (1989) resolved malto-oligosaccharides up to DP 80 on a CarboPac PA1 column using a high-pH mobile phase, therefore, it is possible to resolve a DP 50 oligosaccharide with this column. Furthermore, Thibault et al. (1993) showed that apple, beet and citrus pectin, following partial acid hydrolysis, contained homogalacturonans (98 + % galacturonic acid content) with average molecular weights equivalent to DP 72–100 oligogalacturonic acids. The galacturonic acid-rhamnose molar ratio for the citrus (source of the polygalacturonic acid used by Hotchkiss and Hicks 1990) homogalacturonan was 247:1 (Thibault et al. 1993). Therefore, it is extremely likely that the oxalate gradient HPAEC-PAD peaks represent a homopolymer series of individually resolved oligogalacturonic acids. We correlated the average radius of gyration values calculated from GPC chromatograms with those values calculated from oxalate gradient HPAEC-PAD chromatograms of partially hydrolyzed polygalacturonic acid (Hotchkiss et al. 1995a). The smallest pectic macromolecular component was equivalent to the series of oligogalacturonic acids up to DP 40. Therefore, with the oxalate gradient mobile phase, it is posible to bridge the pectin structural information obtained from GPC with that obtained from high-pH HPAEC-PAD.

Quantitation of the carbohydrate composition by HPAEC-PAD is not a straightforward process. There has been some debate concerning the correlation between oligosaccharide molecular weight and PAD response. A better correlation does exist between carbohydrate molar concentration and PAD response than for carbohydrate mass concentration (Paskach et al. 1991). However, this correlation is not as linear as that obtained from refractive index or UV-absorbance HPLC detectors. Assuming the PAD detects on a molar basis when different carbohydrates with different chemical and physical properties are analyzed can lead to error. Clearly, when considering a homopolymer series of oligosaccharides, the PAD molar response is not linear with increasing molecular weight (Hotchkiss and Hicks 1990; Koizumi et al. 1991a; Mort et al. 1993). The change in relative PAD response with increasing DP of galacturonic acid oligosaccharides terminated by galactose at their reducing ends was linear up to DP 7 (Mort et al. 1993). However, PAD quantitation progressively underestimated the amounts of these oligosaccharides as the DP increased compared to UV detection following labeling with 2-aminopyridine. The latter labeling and detection method provides equimolar response for oligogalacturonic acids up to DP 20 (Maness and Mort 1989). With malto-oligosaccharides, where the increase in PAD molar response was reported to be proportional to the number of sugar ring hydroxyl groups as molecular weight increased (Koizumi et al. 1989), the molar response is nonlinear above DP 10 (Koizumi et al. 1991a). Therefore, HPAEC-PAD is only quantitative if exact standards for the analyte of interest are available. HPAEC-MS shows great potential for the identification of unknown HPEAC-PAD peaks when standards are not available. However, HPAEC-MS is currently limited (by the desalting method prior to the thermospray interface) to oligosaccharides not larger than a DP 3 oligogalacturonic acid or DP 9 rhamnogalacturonan oligosaccharide (Schols et al. 1994a). Detection based on mass was demonstrated for HPAEC-separated oligogalacturonic acids up to DP 15 using a continuous postcolumn permanganate reaction system and absorbance detection at 525 nm (Thomas and Mort 1994). Mass basis detection of oligogalacturonic acids avoids the loss of sensitivity from the PAD response, or from the molar response of reducing end labeled oligosaccharides that is associated with increasing DP length. However, it should be noted that the CarboPac HPAEC column still has unequaled selectivity for oligosaccharide separations, which makes up for disadvantages associated with HPAEC-PAD.

3 Pectin Analysis in Ripening Peach Fruit

3.1 Melting- and Nonmelting-Flesh Peaches

Many examples have been reported where endo-polygalacturonase (endo-PG) is associated with pectin depolymerization during fruit ripening (Brady 1987). Higher endo-PG activity was proposed to be responsible for the increased pectin solubilization and softening in melting-flesh (freestone) peaches relative to nonmelting-flesh (clingstone) peaches (Pressey and Avants 1978). Due to the rapid decrease in tissue firmness (late during ripening) in melting-flesh peaches compared to nonmelting-flesh peaches, the former peaches are used for the fresh market and the latter used for canning (Lester et al. 1994). Melting-flesh peaches contain both exo- and endo-polygalacturonase enzymes, while nonmelting flesh peaches only have exo-polygalacturonase activity during ripening (Pressey 1986). Endo-PG gene expression, probed with a partial cDNA sequence from ripe melting-flesh peach fruit, occurred in low levels as the fruit began to ripen and high levels in ripe fruit (Lester et al. 1994). Low levels of a different RNA transcript were detected with this probe only in ripe nonmelting-flesh fruit (Lester et al. 1994). Three pectin methyl-esterase (PME) isozymes have been isolated from ripe melting-flesh peaches (Glover and Brady 1994). In tomato, the relationship between endo-PG activity and fruit ripening has become unclear because while expression of anti-sense RNA to this enzyme prevented pectin degradation, the tissue still softened (Sheehy et al. 1988). These observations led to an investigation of other carbohydrate-modifying enzyme activities that may play a role in tomato ripening (Maclachlan and Brady 1994).

Fishman et al. (1993b) demonstrated that pectin solubilization and softening were greatest in Redskin (melting-flesh) peaches between 21 weeks after flowering (WAF) and 22 WAF compared to Suncling (nonmelting flesh) peaches at the same stage of ripening. The size and intrinsic viscosity of the Redskin chelator-soluble pectin decreased and the molecular weight of both chelator-soluble and alkaline-soluble Redskin pectin increased between 21 and 22 WAF compared to Suncling pectin (Fishman et al. 1993b). Therefore, endo-PG is only active at the end of the melting-flesh peach ripening process when the most dramatic textural softening occurs. These authors suggested that other mechanisms may be involved in the initial stages of Redskin peach ripening and Suncling peach softening (Fishman et al. 1993b). Redskin peach pectin (20 WAF alkaline-soluble pectin) consists of four macromolecular components in which the larger components are aggregates of the smaller components (Fishman et al. 1992, 1993a). All but the smallest component could be visualized by electron microscopy as a network of segmented, kinked, and branched rods which had a discontinuous distribution of contour lengths.

3.2 HPAEC-PAD System

Conditions for HPAEC-PAD were similar to those reported previously (Hotchkiss and Hicks 1990). HPAEC-PAD was performed with a Dionex Bio-LC system, which included a PAD 2 (gold working electrode) pulsed amperometric detector (E_1 = 0.15 V, 480 ms; E_2 = 0.7 V, 120 ms; E_3 = −0.6 V, 360 ms), and a pressurized bottle post-column delivery system that added 500 mM potassium hydroxide prior to the detector. A CarboPac PA1 column (4 × 250 mm; Dionex Corp.) and a CarboPac PA Guard column (3 × 25 mm; Dionex Corp.) were utilized. The mobile phase consisted of a 25–500 mM potassium oxalate pH 6 nonlinear, 100-min mobile phase gradient (Table 1).

Table 1. HPAEC-PAD mobile phase gradient

Time (min)	Eluant A 500 mM K^+ oxalate (%)	Elunat B H_2O (%)	Oxalate (mM)
0–1	5	95	25
9	20	80	100
40	40	60	200
65	50	50	250
80	53	47	265
100	100	0	500
110	100	0	500
111	5	95	25

3.3 Sample Preparation

Peach cell walls were isolated and extracted sequentially with a chelating agent (50 mM CDTA) and then under alkaline conditions (50 mM Na_2CO_3, 2 mM CDTA) as reported previously (Fishman et al. 1993b), producing chelator-soluble pectin (CSP) and alkaline-soluble pectin (ASP), respectively. CSP and ASP were hydrolyzed by incubating 10 mg of pectin in 0.1 M NaCl, pH 4, 30 °C to which 1 unit of purified endo-polygalacturonase (Doner et al. 1988) was added (1 ml total volume). Hydrolysis was terminated by raising the pH to 6 with KOH and heating the sample in a boiling water bath for 10 min. A 200-µl aliquot was removed, filtered (0.45 µm), and injected for HPAEC-PAD analysis. The pH of the remaining sample was adjusted to 4 with CH_3COOH and hydrolysis continued (30 °C) after adding an additional unit of endo-polygalacturonase. Due to a limited amount of sample available for analysis, this process was repeated for each time point during the hydrolysis time course.

3.4 HPAEC-PAD of Redskin and Suncling Peach Pectin

HPAEC-PAD analysis of Redskin and Suncling peach pectins revealed only trace amounts (if any) of oligogalacturonic acids and a broad, occasionally bimodal, peak at approximately 90 min that we have described as a "polymer" peak (Hotchkiss and Fishman 1992). The polymer peak represents the beginning of the size range at which the CarboPac PA1 column no longer resolves individual oligogalacturonic acids and/or the size range at which the forces that aggregate the larger pectic subunits are stronger than the dissociating properties of the potassium oxalate buffer. The earlier-eluting components of the bimodal polymer peak appears to consist of homogalacturonan since it is depolymerized fastest by endo-PG (Figs. 1a, 2a,b, 3a). However, endo-PG hydrolysis of the polymer peak proceeds by a two-stage reaction. The later-eluting second components of the polymer peak, that which is more slowly depolymerized, is more difficult to characterize. The second polymer peak component probably does not result solely from a higher degree of branching or degree of esterification, which are characteristics of pectins

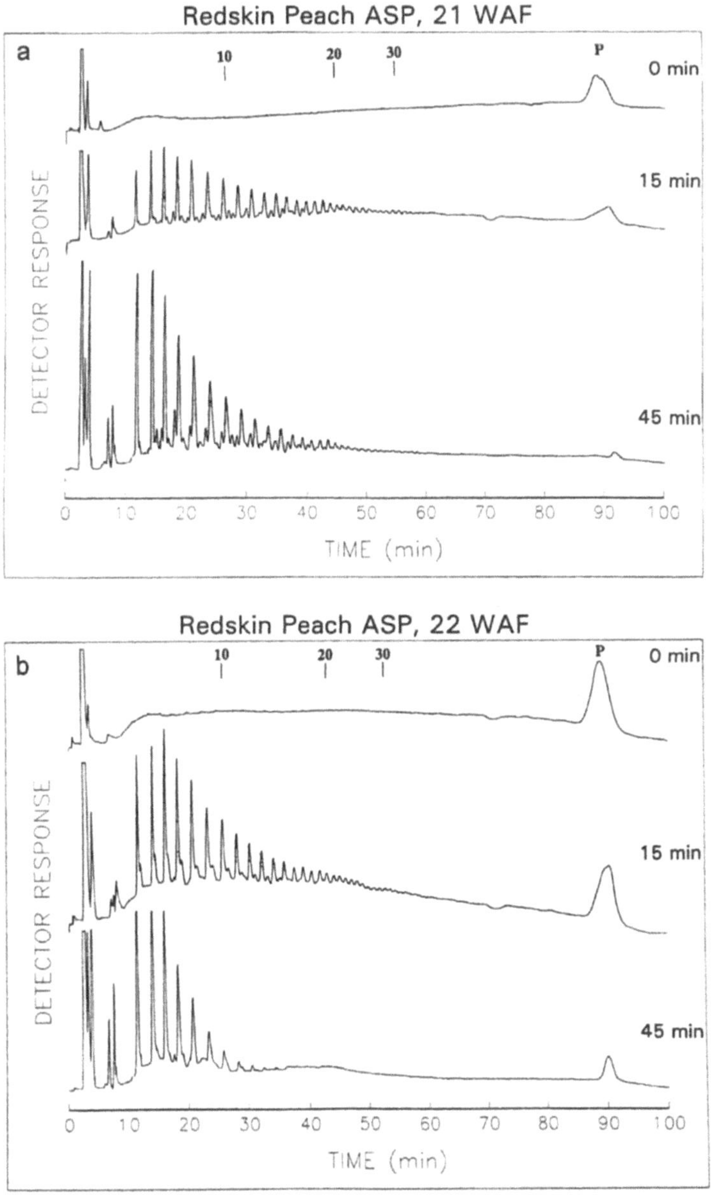

Fig. 1a,b. HPAEC-PAD chromatograms of endo-PG depolymerization of Redskin peach ASP at **a** 21 WAF and **b** 22 WAF. The time after addition of the enzyme is indicated at the end of each chromatogram. The *numbers* over each chromatogram refer to the oligogalacturonic acid DP determined by comparison of retention times with standards (Doner et al. 1988) and then extrapolation; *P* polymer peak

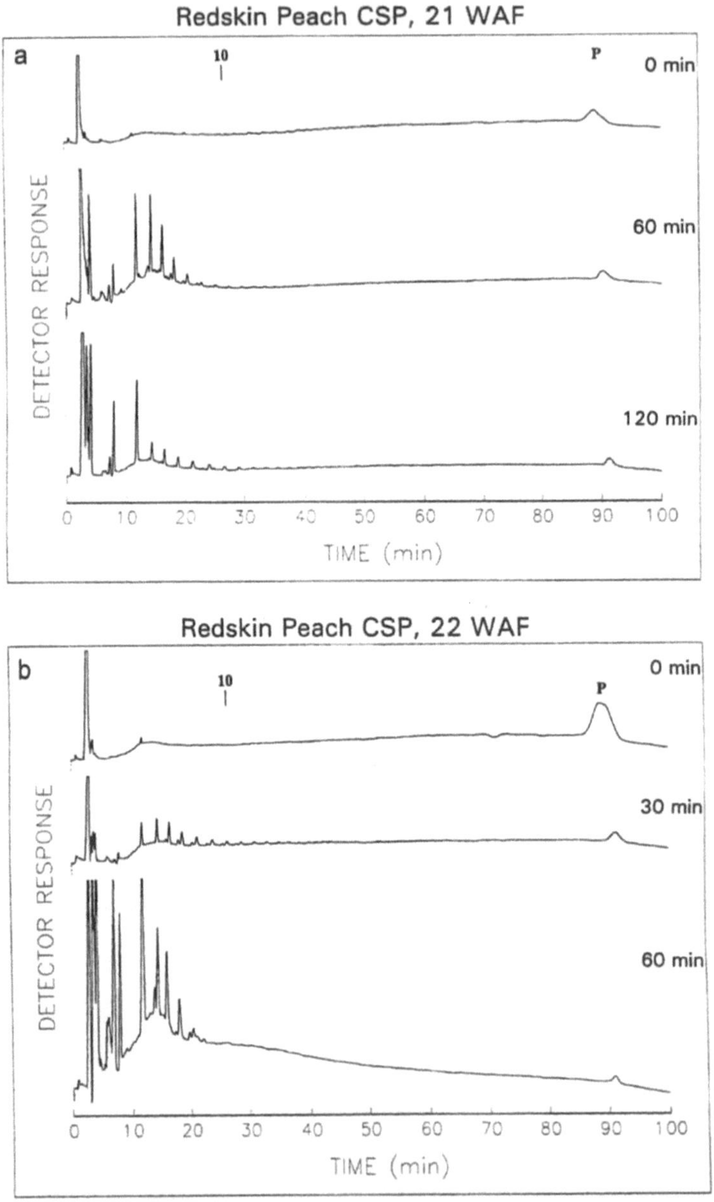

Fig. 2a,b. Endo-PG depolymerization of Redskin CSP at **a** 21 WAF and **b** 22 WAF. See Fig. 1 for description of labels

extracted under alkaline conditions. Increases in both of these structural properties in oligosaccharides leads to earlier elution from the CarboPac PA1 column (Mutter et al. 1994; Schols et al. 1994b). Possibly, a combination of branching, methyl-esterification, and pectin component aggregation may sequester endo-PG susceptible linkages in this later-eluting polymer peak component.

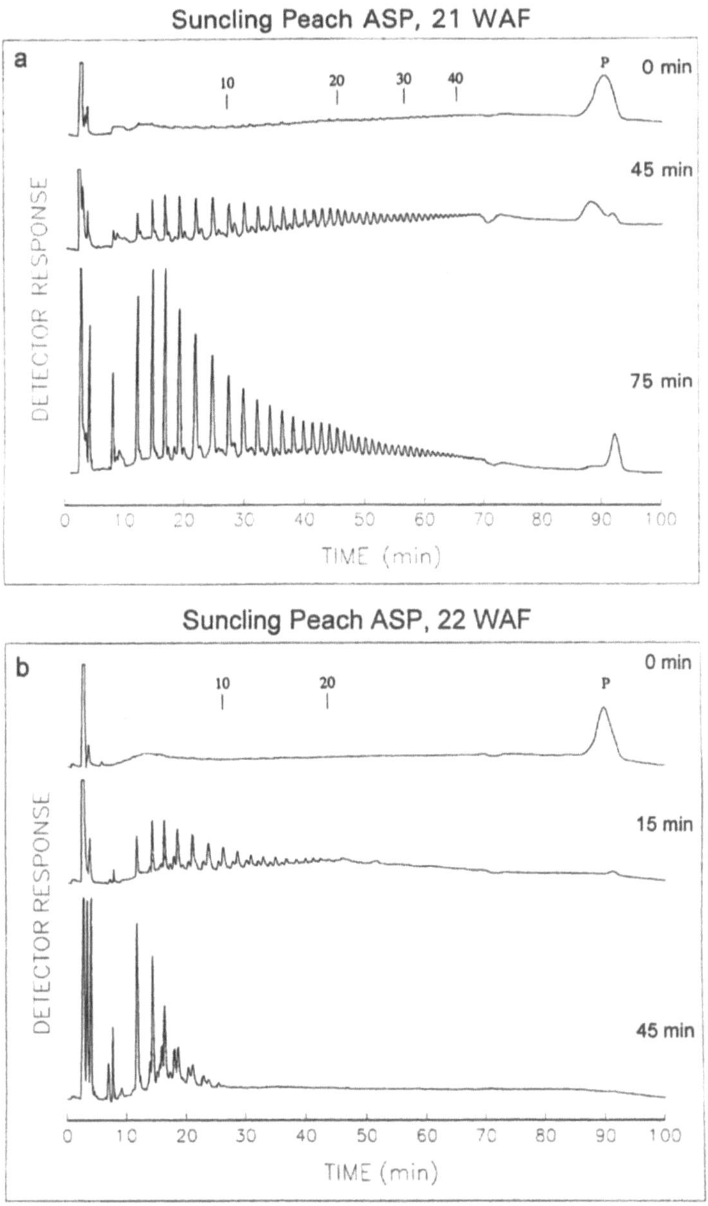

Fig. 3a,b. Endo-PG depolymerization of Suncling peach ASP at **a** 21 WAF and **b** WAF. See Fig. 1 for description of labels

We previously demonstrated that a series of oligogalacturonic acids were generated from 20 WAF Redskin peach ASP following a 30-min treatment with endo-PG (Fishman et al. 1993a). Endo-PG treatment of Redskin peach pectin at 21 and 22 WAF also generated a series of oligogalacturonic acids. However, ASP was generally more susceptible to endo-PG hydrolysis than CSP based on the time required to hydrolyze oligogalacturonic acids. The alkaline conditions, used to extract ASP, saponified methyl-ester groups, rendering more pectin available for hydrolysis. Methyl-esterification of pectin (Thibault and Mercier 1978) and acetyl-esterification of galacturonic acid at C-2 and C-3 (Rexova-Benkova et al. 1977) decreases the extent of pectin degradation by endo-PG. The difference between the susceptibility of ASP and CSP to endo-PG is much more noticeable at 21 WAF than at 22 WAF, which correlates with changes in radius of gyration and intrinsic viscosity observed at this stage of ripening. The radius of gyration and intrinsic viscosity of CSP is much higher than that of ASP at 21 WAF (Fishman et al. 1993b). However, by 22 WAF these parameters for CSP and ASP are almost identical (Fishman et al. 1993b). Additionally, larger oligogalacturonic acids were released from ASP by endo-PG hydrolysis relative to CSP. While de-esterification with alkali occurs in a random pattern (Rexova-Benkova and Markovic 1976), larger blocks of homogalacturonan may have been produced by the alkaline extraction conditions in the ASP compared to that present in the CSP.

An increase in susceptibility of the Redskin CSP to endo-PG hydrolysis between 21 and 22 WAF is coincident with the significant decreases in radius of gyration and intrinsic viscosity, and increase in molecular weight reported earlier (Fishman et al. 1993b). Endo-PG-mediated hydrolysis of long, thin CSP aggregates (smooth regions) during this stage of ripening leaving short thick pectin aggregates (hairy regions) was proposed to be the major pectin structural difference between melting-flesh and nonmelting-flesh peach softening (Fishman et al. 1993b). Our results reflect the action of PME or acetyl-esterase during this stage of Redskin peach ripening. These enzymes deesterify CSP such that the 22 WAF CSP is more susceptible to endo-PG hydrolysis than it was at 21 WAF (Fig. 2). A higher CSP degree of esterification at 21 WAF could also be responsible for the higher radius of gyration and intrinsic viscosity values observed since it would be less able to aggregate due to calcium cross-linking. Chilling injury of peach fruit during storage that leads to a mealy texture called "woolliness", has been explained by an imbalance between PME and endo-PG activities such that high molecular weight, low-methoxy pectins accumulate (Ben-Arie and Sonego 1980). The enrichment in hairy regions in the 22 WAF Redskin CSP compared to 21 WAF is illustrated by the increased number of secondary peaks adjacent to the oligogalacturonic acid peaks (more minor peaks and shoulders in Fig. 2b, 60 min compared to Fig. 2a, 60 min). These secondary peaks are produced by oligogalacturonic acids containing one or more neutral sugars such as rhamnose. These secondary peaks are also present in Redskin ASP hydrolysates (Fig. 1; see also Fig. 10 in Fishman et al. 1993a).

HPAEC-PAD analysis of Suncling peach pectin (Figs. 3 and 4) demonstrated trends similar to those observed with Redskin peach pectin. ASP was more susceptible to endo-PG hydrolysis than CSP, and 22 WAF pectin (ASP and CSP) was more susceptible than 21 WAF pectin. Therefore, while endo-PG does not play a role in the normal ripening of this nonmelting-flesh peach, the Suncling pectin is just as susceptible to endo-PG hydrolysis at 21 WAF as Redskin pectin. The increase in endo-PG susceptibility between 21 and 22 WAF suggests that *in muro* esterase activity may play a role in Suncling peach softening. Oligogalacturonic acids re-

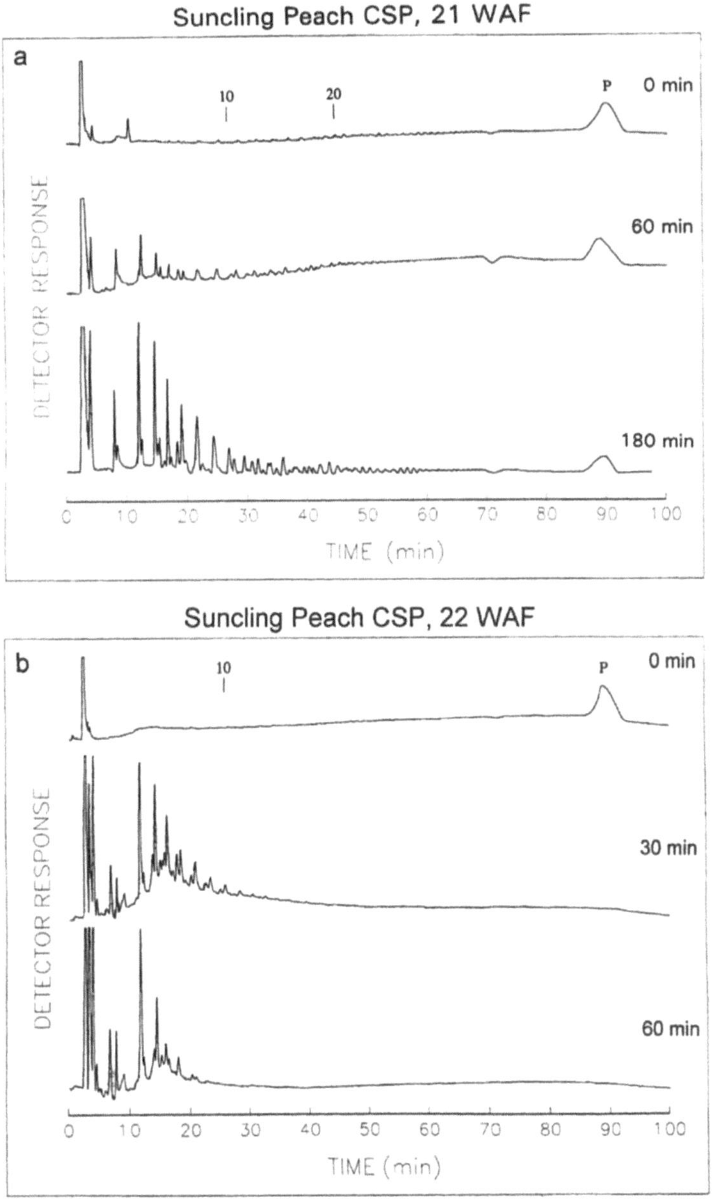

Fig. 4a,b. Endo-PG depolymerization of Suncling peach CSP at **a** 21 WAF and **b** 22 WAF. See Fig. 1 for description of labels

leased from Suncling pectin were consistently larger than those released from Redskin pectin during endo-PG hydrolysis. This suggests that the homogalacturonan DP size is larger in Suncling pectin, which probably reflects the relatively low levels of endo-PG activity during ripening.

4 Conclusions and Future Directions

HPAEC-PAD analysis of peach pectin and the susceptibility of these polysaccharides to endo-polygalacturonase hydrolysis provided valuable insights concerning the ripening process of these fruits. The degree of pectin esterification was implicated as the molecular control of peach softening since ASP as more susceptible to endo-PG hydrolysis than CSP and susceptibility increased between 21 and 22 WAF. Other carbohydrate-modifying enzymes, such as PME, may play a significant role in the initial stages of melting-flesh peach ripening. PME also appears to play a role in nonmelting-flesh peach ripening. The larger size range of oligogalacturonic acids released from nonmelting flesh peach pectin compared to melting-flesh peach pectin confirmed the previously reported role of endo-PG in the softening of the latter peaches. Through comparison with the HPSEC data reported for these pectins, meaningful interpretations of the HPAEC-PAD data could be made. Homogalacturonan and an aggregation of larger pectic components were the two parts of the polymer peak represented in HPAEC-PAD chromatograms. The complete picture of changes in pectin structure is only drawn when both levels of analytical resolution are applied.

The development of CarboPac PA columns advanced dramatically the field of oligosaccharide separations. New stationary phases consisting of a highly cross-linked polystyrene-divinylbenzene surface quaternized with amino functions (Corradini et al. 1994), amine surface-functionalized nonporous microspherical zirconia (Yu and Elrassi 1994), DEAE-functionalized polystyrene-divinylbenzene (301 VHP, Vydac; GlycoSep C, Oxford Glycosystems), porous graphitized carbon (Koizumi et al. 1991b; Hypercarb, Shandon; GlycoSep H, Oxford Glycosystems), aminopropylsilica gel covalently coated with polysuccinimide (Alpert 1990; Alpert et al. 1994; PolyHydroxyethyl A and PolyGLYCOPLEX, PolyLC), and β-cyclodextrin bonded silica gel (Simms et al. 1995; Cyclobond I, Advanced Separations Technologies) may expand oligosaccharide separation selectivity further in the future. Additionally, capillary zone electrophoresis has great potential for pectic oligosaccharide separations (Liu et al. 1992).

References

Alpert A (1990) Hydrophilic-interaction chromatography for the separation of peptides, nucleic acids and other polar compounds. J Chromatogr 499:177–196

Alpert A, Shukla M, Shukla AK, Zieske LR, Yuen SW, Ferguson MAJ, Mehlert A, Pauly M, Orlando R (1994) Hydrophilic-interaction chromatography of complex carbohydrates. J Chromatogr 676:191–202

An J, O'Neill MA, Albersheim P, Darvill AG (1994a) Isolation and structural characterization of β-D-glucosyluronic acid and 4-O-methyl β-D-glucosyluronic acid-containing oligosaccharides from the cell-wall pectic polysaccharide, rhamnogalacturonan I. Carbohydr Res 252:235–243

An J, Zhang L, O'Neill MA, Albersheim P, Darvill AG (1994b) Isolation and structural characterization of endo-rhamnogalacturonase-generated fragments of the backbone of rhamnogalacturnoan I. Carbohydr Res 264:83–96
Barth HG (1980) High-performance gel permeation chromatography of pectins. J Liq Chromatogr 3:1481–1496
Ben Arie R, Sonego L (1980) Pectolytic enzyme activity involved in woolly breakdown of stored peaches. Phytochemistry 19:2553–2555
Brady C (1987) Fruit ripening. Annu Rev Plant Physiol 38:155–178
Colquhoun IJ, De Ruiter GA, Schols HA, Voragen AGJ (1990) Identification by n.m.r. spectroscopy of oligosaccharides obtained by treatment of the hairy regions of apple pectin with rhamnogalacturonase. Carbohydr Res 206:131–144
Corradini C, Corradini D, Huber CG, Bonn GK (1994) Synthesis of a polymeric-based stationary phase for carbohydrate separation by high-pH anion-exchange chronmatography with pulsed-amperometric detection. J Chromatogr 685:213–220
Deckers HA, Olieman C, Rombouts FM, Pilnik W (1986) Calibration and application of high-performance size exclusion columns for molecular weight distribution of pectins. Carbohydr Polymers 6:361–378
Derungs H, Deuel H (1954) Quantitative Trennung von Oligogalakturonsäuren an Ionenaustauschern. Helv Chim Acta 37:657–659
De Vries JA, Voragen AGJ, Rombouts FM, Pilnik W (1981) Extraction and purification of pectins from alcohol insoluble solids from ripe and unripe apples. Carbohydr Polymers 1:117–127
De Vries JA, Rombouts FM, Voragen AGJ, Pilnik W (1982) Enzymic degradation of apple pectins. Carbohydr Polymers 2:25–33
Doner LW, Irwin PL, Kurantz MJ (1988) Preparative chromatography of oligogalacturonic acids. J Chromatogr 449:229–239
Endress H-U, Omran H, Gierschner K (1991) Monitoring the course of enzymic degradation of pectic substances by automated fast ion chromatography (FIC). Lebensm Wiss Technol 24:80–85
Fishman ML, Pfeffer PE, Barford RA, Doner LW (1984) Studies of pectin solution properties by high-performance size exclusion chromatography. J Agric Food Chem 32:372–378
Fishman ML, Damert WC, Phillips JG, Barford RA (1987) Evaluation of root-mean-square radius of gyration as a parameter for universal calibration of polysaccharides. Carbohydr Res 160:215–225
Fishman ML, Gillespie DT, Sondey SM, Barford RA (1989a) Characterization of pectins by size exclusion chromatography in conjunction with viscosity detection. J Agric Food Chem 37:584–591
Fishman ML, Gross KC, Gillespie DT, Sondey SM (1989b) Macromolecular components of tomato fruit pectin. Arch Biochem Biophys 274:179–191
Fishman ML, Cooke P, Levaj B, Gillespie DT, Sondey SM, Scorza R (1992) Pectin microgels and their subunit structure. Arch Biochem Biophys 294:253–260
Fishman ML, Cooke P, Hotchkiss A, Damert W (1993a) Progressive dissociation of pectin, Carbohydr Res 248:303–316
Fishman ML, Levaj B, Gillespie DT, Scorza R (1993b) Changes in the physico-chemical properties of peach fruit pectin during on-tree ripening and storage. J Am Soc Hortic Sci 118:343–349
Fry SC, Aldington S, Hetherington PR, Aitken J (1993) Oligosaccharides as signals and substrates in the plant cell wall. Plant Physiol 103:1–5
Glover H, Brady C (1994) Purification of three pectin esterases from ripe peach fruit. Phytochemistry 37:949–955
Hardy MR, Townsend RR (1988) Separation of positional isomers of oligosaccharides and glycopeptides by high-performance anion-exchange chromatography with pulsed amperometric detection. Proc Natl Acad Sci USA 85:3289–3293
Heyraud A, Rochas C (1982) Sulfated oligosaccharides from k-carrageenan and oligogalacturonic acids separation by H.P.L.C. J Liq Chromatogr 5:403–412
Hotchkiss AT Jr, Fishman ML (1992) Pectin structure revealed by HPAE-PAD. Plant Physiol 99:46a
Hotchkiss AT Jr, Hicks KB (1989) Analysis of oligogalacturonic acids by high performance anion exchange chromatography with pulsed amperometric detection. Plant Physiol 89:43a

Hotchkiss AT Jr, Hicks KB (1990) Analysis of oligogalacturonic acids with 50 or fewer residues by high-performance anion-exchange chromatography and pulsed amperometric detection. Anal Biochem 184:200–206

Hotchkiss AT Jr, Hicks KB (1993) Analysis of pectate lyase-generated oligogalacturonic acids by high-performance anion-exchange chromatography with pulsed amperometric detection. Carbohydr Res 247:1–7

Hotchkiss AT Jr, Hicks KB, Orlandi EW, Baker CJ (1990) Simultaneous analysis of oligogalacturonic acids with and without a 4,5-unsaturated function at the nonreducing end. Plant Physiol 93:76a

Hotchkiss AT Jr, Hicks KB, Doner LW, Irwin PL (1991) Isolation of oligogalacturonic acids in gram quantities by preparative h.p.l.c. Carbohydr Res 215:81–90

Hotchkiss At Jr, Fishman JL, Hoagland PD (1995a) A comparative analysis of pectic macromolecular components by HPSEC and HPAEC-PAD. Carbohydr Polymers (submitted)

Hotchkiss AT Jr, Revear LG, Hicks KB (1995b) Substrate depolymerization pattern of *Pseudomonas viridiflava* SF-312 pectate lyase. Physiol Mol Plant Pathol (in press)

Huber DJ, O'Donoghue EM (1993) Polyuronides in avocado (*Persea americana*) and tomato (*Lycopersicon esculentum*) fruits exhibit markedly different patterns of molecular weight downshifts during ripening. Plant Physiol 102:473–480

Jarvis MC (1984) Structure and properties of pectin gels in plant cell walls. Plant Cell Environ 7:153–164

Kofod LV, Kauppinen S, Christgau S, Andersen LN, Heldt-Hansen HP, Dörreich K, Dalboge H (1994) Cloning and characterization of two structurally and functionally divergent rhamnogalacturonases from *Aspergillus aculeatus*. J Biol Chem 269:29182–29189

Koizumi K, Kubota Y, Tanimoto T, Okada Y (1989) High-performance anion-exchange chromatography of homogeneous D-gluco-oligosaccharides and -polysaccharides (polymerization degree ≥50) with pulsed amperometric detection. J Chromatogr 464:365–373

Koizumi K, Fukuda M, Hizukuri S (1991a) Estimation of the distributions of chain length of amylopectins by high-performance liquid chromatography with pulsed amperometric detection. J Chromatogr 585:233–238

Koizumi K, Okada Y, Fukuda M (1991b) High-performance liquid chromatography of mono- and oligo-saccharides on a graphitized carbon column. Carbohydr Res 215:67–80

Lau JM, McNeil M, Darvill AG, Albersheim P (1985) Structure of the backbone of rhamnogalacturonan I, a pectic polysaccharide in the primary cell walls of plants. Carbohydr Res 137:111–125

Lee YC (1990) High-performance anion-exchange chromatography for carbohydrate analysis. Anal Biochem 189:151–162

Lieker H-P, Thielecke K, Buchholz K, Reilly PJ (1993) High-performance anion-exchange chromatography of saturated and unsaturated oligogalacturonic acids. Carbohydr Res 238:307–311

Lerouge P, O'Neill MA, Darvill AG, Albersheim P (1993) Structural characterization of endo-glycanase-generated oligoglycosyl side chains of rhamongalacturonan I. Carbohydr Res 243:359–371

Lester DR, Speirs J, Orr G, Brady CJ (1994) Peach (*Prunus persica*) endopolygalacturonase cDNA isolation and mRNA analysis in melting and nonmelting peach cultivars. Plant Physiol 105:225–231

Liu J, Shirota O, Novotny MV (1992) Sensitive, laser-assisted determination of complex oligosaccharide mixtures separated by capillary gel electrophoresis at high resolution. Anal Chem 64:973–975

Lo V-M, Hahn MG, Van Halbeek H (1994) Preparation, purification, and structural characterization of linear oligogalacturonides. An FAB-mass spectrometric and NMR spectroscopic study. Carbohydr Res 255:271–284

MacDougall AJ, Rigby NM, Needs PW, Selvendran RR (1992) Movement and metabolism of oligogalacturonide elicitors in tomato shoots. Planta 188:566–574

Maclachlan G, Brady C (1994) Endo-1,4-β-glucanase, xyloglucanase, and xyloglucan endo-transglycosylase activities versus potential substrates in ripening tomatoes. Plant Physiol 105:965–974

Maness NO, Mort AJ (1989) Separation and quantitation of galacturonic acid oligomers from 3 to over 25 residues in length by anion-exchange high-performance liquid chromatography. Anal Biochem 178:248–254

McCann MC, Wells B, Roberts K (1992) Complexity in the spatial localization and length distribution of plant cell-wall matrix polysaccharides. J Microsc 166:123–136

McNeil M, Darvill AG, Fry SC, Albersheim P (1984) Structure and function of the primary cell walls of plants. Annu Rev Biochem 53:625–663

Melotto E, Greve LC, Labavitch JM (1994) Cell wall metabolism in ripening fruit. VII. Biologically active pectin oligomers in ripening tomato (*Lycopersicon esculentum* Mill.) fruits. Plant Physiol 106:575–581

Mort AJ, Qiu F, Maness NO (1993) Determination of the pattern of methyl esterification in pectin. Distribution of contiguous nonesterified residues. Carbohydr Res 247:21–35

Mutter M, Beldman G, Schols HA, Voragen AGJ (1994) Rhamnogalacturonan α-L-Rhamnopyranohydrolase. A novel enzyme specific for the terminal nonreducing rhamnosyl unit in rhamnogalacturonan regions of pectin. Plant Physiol 106:241–250

Mutter M, Colquhoun IJ, Beldman, G, Schols HA, Voragen AGJ (1996) Rhamnogalacturonan α-L-rhamnopyranosyl-(1,4)-galactopyranosyluronide lyase. Plant Physiol (in press)

Neidhart S, Gessler A, Hannak C, Endress H-U, Gierschner K (1994) Anwendung der analytischen Grossenausschluss-Chromatographie zur Untersuchung des Substratabbaus einer kommerziellen Pektatlyase aus *Erwinia caratovora.* Z Lebensm Unters Forsch 198:317–324

Nothnagel EA, McNeil M, Albersheim P, Dell A (1983) Host-pathogen interactions. XXII. A galacturonic acid oligosaccharide from plant cell walls elicits phytoalexins. Plant Physiol 71:916–926

O'Neill M, Albersheim P, Darvill AG (1990) The pectic polysaccharides of primary cell walls. In: Dey PM (ed) Methods in plant biochemistry. Academic Press, New York, pp 415–441

Paskach TJ, Lieker H-P, Reilly PJ, Thielecke K (1991) High-performance anion-exchange chromatography of sugars and sugar alcohols on quaternary ammonium resins under alkaline conditions. Carbohydr Res 215:1–14

Powell DA, Morris ER, Gidley MJ, Rees DA (1982) Conformations and interactions of pectins. II. Influence of residue sequence on chain association in calcium pectate gels. J Mol Biol 155:517–531

Pressey R (1986) Polygalacturonases in higher plants. In: Fishman ML, Jen JJ (eds) Chemistry and function of pectins. American Chemical Society, Washington, DC, pp 157–174

Pressey R (1991) Oxidized oligogalacturonides activate the oxidation of indoleacetic acid by peroxidase. Plant Physiol 96:1167–1170

Pressey R, Avants JK (1978) Differences in polygalacturonase composition of clingstone and freestone peaches. J Food Sci 43:1415–1423

Preston JF, Rice JD (1991) Kinetic analysis of pectate lyases by high-performance liquid chromatography. Carbohydr Res 215:137–145

Rexová-Benková L (1970) Separation of oligogalacturonic acids by dextran gel chromatography. Chem Zvesti 24:59–62

Rexová-Benková L, Markovic O (1976) Pectic enzymes. Adv Carbohydr Chem Biochem 33:323–385

Rexová-Benková L, Mracková M, Luknár O, Kohn R (1977) The role of sec-alcohol groups of D-galacturonan in its degradation by endo-D-galacturonanase. Collect Czech Chem Commun 42:3204–3213

Rouau X, Thibault J-F (1984) Apple juice pectic substances. Carbohydr Polymers 4:111–125

Schols HA, Voragen AGJ (1994) Occurrence of pectic hairy regions in various plant cell wall materials and their degradability by rhamnogalacturonase. Carbohydr Res 256:83–95

Schols HA, Reitsma JCE, Voragen AGJ, Pilnik W (1989) High-performance ion exchange chromatography of pectins. Food Hydrocolloids 3:115–121

Schols HA, Ceraeds CCJM, Searle-van Leeuwen MF, Kormelink FJM, Voragen AGJ (1990a) Rhamnogalacturonase: a novel enzyme that degrades the hairy regions of pectins. Carbohydr Res 206:105–115

Schols HA, Posthumus MA, Voragen AGJ (1990b) Structural features of hairy regions of pectins isolated from apple juice produced by the liquefaction process. Carbohydr Res 206:117–129

Schols HA, Mutter M, Voragen AGJ, Niessen WMA, Van der Hoeven RAM, Van der Greef J, Bruggink C (1994a) The use of combined high-performance anion-exchange chromatography-thermospray mass spectrometry in the structural analysis of pectic oligosaccharides. Carbohydr Res 261:335–342

Schols HA, Voragen AGJ, Colquhoun IJ (1994b) Isolation and chracterization of rhamnogalacturonan oligomers, liberated during degradation of pectic hairy regions by rhamnogalacturonase. Carbohydr Res 256:97–111

Sheehy RE, Kramer MK, Hiatt WR (1988) Reduction of polygalacturonase activity in tomato fruit by antisense RNA. Proc Natl Acad Sci USA 85:8805–8809

Simms PJ, Hotchkiss AT Jr, Irwin PL, Hicks KB (1995) High performance liquid chromatographic separation of oligogalacturonic acids on a β-cyclodextrin bonded phase column. Carbohydr Res (in press)

Spiro MD, Kates KA, Koller AL, O'Neill MA, Albersheim P, Darvill AG (1993) Purification and characterization of biologicall active 1,4-linked α-D-oligogalacturonides after partial digestion of polygalacturonic acid with endopolygalacturonase. Carbohydr Res 247:9–20

Thibault J-F (1980) Separation of α-D-galacturonic acid oligomers by chromatography on polyacrylamide gel. J Chromatogr 194:315–322

Thibault J-F, Mercier C (1978) *Aspergillus niger* endopolygalacturonase. 2. Characterization and some properties. J Food Biochem 2:379–393

Thibault J-F, Renard CMGC, Axelos MAV, Roger P, Crepeau M-J (1993) Studies of the length of homogalacturonic regions in pectins by acid hydrolysis. Carbohydr Res 238:271–286

Thomas J, Mort AJ (1994) Continuous postcolumn detection of underivatized polysaccharides in high-performance liquid chromatography by reaction with permanganate. Anal Biochem 223:99–104

Voragen AGJ, Schols HA, De Vries JA, Pilnik W (1982) High-performance liquid chromatographic analysis of uronic acids and oligogalacturonic acids. J Chromatogr 244:327–336

Voragen AGJ, Schols HA, Gruppen H (1993) Structural studies of plant cell-wall polysaccharides using enzymes. In: Meuser F, Manners DJ, Seibel W (eds) Plant polymeric carbohydrates. Royal Society of Chemistry, Cambridge, pp 3–15

Yu J, Elrassi Z (1994) Preparation of amino-zirconia bonded phases and their evaluation in hydrophilic interaction chromatography of carbohydrates with pulsed amperometric detection. J High Resolution Chromatogr 17:773–778

Characterization of Oligosaccharides Derived from Plant Cell Wall Polysaccharides by On-Line High-Performance Anion-Exchange Chromatography Thermospray Mass Spectrometry

W.M.A. Niessen, H.A. Schols, R.A.M. van der Hoeven, A.G.J. Voragen, and J. van der Greef

1 Introduction

Polysaccharides have a variety of important biological functions in plant and microbial cell walls or may be used for energy storage in the cell. The polysaccharide-containing plant cell walls are exploited in many ways, e.g. as building material, textiles, fodder and paper, while polysaccharides extracted from plant cell wall material are used food ingredients and as constituents in paint, cosmetics and pharmaceuticals. Such applications rely on the physicochemical (and biological) properties of polysaccharides and actuate research into the structure-function relationship. Since polysaccharides embody the bulk of the plant biomass, versatile and rapid methods for the elucidation of the chemical structure of polysaccharides are required.

The work described here is part of a project aiming at the development of advanced methodology for the structure elucidation of water-extractable as well as water-unextractable plant cell wall materials. For the latter, isolation procedures are performed in which non-cell-wall components like fat, starch, proteins and water-soluble polysaccharides are removed. The water-unextractable cell wall material is then treated with purified enzymes to hydrolyze specific types of sugar linkages in order to release and solubilize oligomeric and polymeric products. Additional specific extraction procedures might be needed to open the cell wall material for enzymic digestion. By studying fragments released from both soluble and insoluble polysaccharides, more knowledge is gained on the structure of the polymer. Successive use of several well-characterized enzymes provides valuable information on the structure as well as on the interaction of various polysaccharides in plant material. The polymeric and oligomeric products released are separated and fractionated by various types of chromatography, e.g. high-performance anion-exchange chromatography (HPAEC) and size-exclusion chromatography. Finally, the fractions are characterized by determination of the glycosidic linkage composition by gas chromatography-mass spectrometry, NMR spectroscopy and fast-atom bombardment (tandem) mass spectrometry (MS) (Reinhold 1986).

For NMR spectroscopy, the extensive, laborious and time-consuming sample preparation including fraction collection and, when HPAEC is used, desalting is

Modern Methods of Plant Analysis, Vol. 17
Plant Cell Wall Analysis
Edited by H.-F. Linskens and J.F. Jackson

inevitable. However, considering the recent developments in interfacing liquid chromatography and mass spectrometry (LC-MS) (Niessen and van der Greef 1992), it is worth investigating the potential of LC-MS in carbohydrate research. On-line LC-MS would greatly facilitate the identification of unknown oligosaccharides without the need for extensive fractionation and purification, especially because well-defined oligosaccharides are generated in the degradation of polysaccharides by specific enzymes. Papers describing the LC-MS analysis of oligosaccharides were surprisingly scarce at the start of our research project in 1990 (Rajakylä 1986; Santikarn et al. 1987; Arpino 1990; Simpson et al. 1990). The development, potential and limitations of the on-line coupling of HPAEC and thermospray MS for the characterization of oligosaccharides enzymically or chemically derived from plant cell wall polysaccharides are described in this chapter. HPAEC is chosen for the coupling to MS because it is currently the most versatile and powerful method for the chromatographic separation of oligosaccharides (Lee 1990; Schols et al., 1994a).

2 Methods

2.1 Apparatus

The general experimental setup was similar to that previously reported by Simpson et al. (1990). The system essentially consists of three building blocks: the anion-exchange chromatograph, the coupling device, and the mass spectrometer equipped with an LC-MS interface. A schematic diagram of the experimental setup in shown in Fig. 1. The anion-exchange chromatography was a Dionex (Sunnyvale, CA, USA) DX-300 modular ion chromatography system, consisting of an EDM-2 solvent degas unit, an AGP pump module, an LCM-3 chromatography module containing a Rheodyne (Cotati, CA, USA) Model 9126 all-PEEK injector with a 25-μl loop, and a pulsed amperometric detector (PAD) with a gold electrode. The chromatograph is coupled to the MS via two anion micromembrane suppressors (AMMS-II) in series, regenerated by a continuous flow of 0.1–0.15 mol/l sulphuric acid at a flow rate of ca. 15 ml/min, and a Kratos (Manchester, UK) Spectroflow 400 LC pump, which acted as a booster pump. The MS was a Finnigan MAT (San José, CA, USA) TSQ-70 tandem mass spectrometer equipped with a Finnigan MAT thermospray interface and a 20-kV conversion-dynode detection system.

Dionex CarboPac PA1 or PA100 columns (250 × 4 mm I.D.) were used in these experiments. The flow rate was 1 ml/min. Gradient elution was applied using two solvents: (A) 0.1 mol/l aqueous sodium hydroxide, and (B) 1 mol/l sodium acetate in 0.1 mol/l aqueous sodium hydroxide. The gradient program used is highly sample-dependent. A number of typical gradient programs used are summarized in Table 1. Each chromatographic run was followed by washing and re-equilibration of the column for 10 min using 100% solvent B and for 10 min using 100% solvent A, prior to setting up the starting conditions of the next gradient. At a T-piece between the AMMS and the booster pump 10^{-4} mol/l aqueous

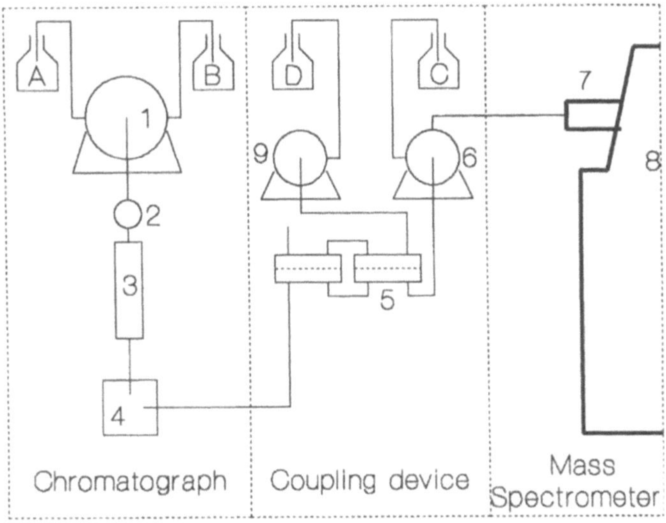

Fig. 1. Schematic diagram of the equipment. *1* Pump; *2* injection valve; *3* column; *4* pulsed amperometric detector; *5* two anion micromembrane suppressors in series; *6* booster pump; *7* thermospray interface; *8* mass spectrometer; *9* regenerant pump. Solvents: *A* 0.1 mol/l aqueous sodium hydroxide; *B* 1 mol/l sodium acetate in aqueous sodium hydroxide; *C* 0.1 mmol/l aqueous sodium acetate; *D* 0.15 mol/l sulphuric acid

Table 1. Typical gradient programs used in HPAEC analysis of oligosaccharides

Sample	Gradient (%B in A)[a]
	CarboPac PA1 column
α-1,4-Glucose	0–30% B in 40 min
α-1,5-Arabinose	0% B for 5 min, 0–40% B in 28 min
β-1,4-Xylose	0–5% B in 5 min, 5–30% B in 40 min
	CarboPac PA100 column
Various (cf. Fig. 2)	0–60% B in 40 min
α-1,4-Glucose	0–30% B in 20 or 30 min
Arabinogalactan-1	0–5% B in 30 min, 5–10% B in 20 min, 10–30% B in 20 min
Arabinogalactan-2	0–20% B in 10 min, 20–25% B in 20 min, 25–50% B in 10 min

[a] Solvent A: 0.1 mol/l sodium hydroxide. Solvent B: 1 mol/l sodium acetate in 0.1 mol/l sodium hydroxide. Flow rate 1 ml/min. Washing and re-equilibration: linear gradient up to 100% B in 5 min, 100% B for 5 min, 100–0% B in 1 min and 15 min at 0% B.

sodium acetate was added at 0.5 ml/min. The total flow rate to the thermospray interface was 1.5 ml/min.

Typical operating conditions of the thermospray interface are a high block temperature of 350 °C in order to reduce ion source contamination by residual sodium salts and sample constituents, a vaporizer temperature of 80–100 °C, and a repeller potential of 50 V. The source chamber, the repeller electrode and the ion sampling cone were rapidly wiped out with wetted cotton wool every 2 or 3 days without taking the source apart.

2.2 Isolation of Oligosaccharides from Plant Cell Wall Material

In order to study the chemical structure of plant cell wall polysaccharides or the oligosaccharides derived from them, the polysaccharides have to be isolated from the starting plant cell wall material without creating artefacts in the original polysaccharide structure. The plant material may also contain fats, proteins or starch, as is for instance the case with soybeans and cereals (Gruppen et al. 1989; Düsterhöft et al. 1991; Verbruggen et al. 1993).

Plant cell wall polysaccharide fractions that are known to consist of specific populations of polysaccharides can be extracted by appropriate extractants. Chelating agents like EDTA and CDTA and low concentrations of alkali at low temperature are frequently used for the extraction of various types of pectin (Selvendran and Ryden 1990). Cell wall material as well as the solubilized polysaccharides can be degraded using highly specific and well-characterized enzymes. The latter are preferred above chemical methods which are much less defined. Characterization of the fragments released by pure enzymes gives valuable information on the structure of the original polysaccharide (Voragen 1993). So far, the characterization of the various oligomers in a digest is performed after laborious isolation steps using, for example, preparative HPAEC. A pre-separation of the enzyme digest using size-exclusion chromatography with a separation range of 200–2000 daltons might be useful to predict the molecular mass ranges of oligosaccharides in the complex mixture. After tedious desalting steps, the structure of the pure oligomers can be determined, in which NMR spectroscopy has proven to be a powerful tool (Gruppen et al. 1992; Schols et al. 1994b).

The arabino-β-1,4-galactan used in this study was isolated from dehulled soybeans. Starting from soy flour, the fat, proteins and starch were removed as described by Verbruggen et al. (1993). A fraction rich in arabinose, galactose and galacturonic acid was obtained by extraction with sodium hydroxide at low temperature. The arabinogalactan fractions obtained were treated by a purified β–1,4-galactanase (van de Vis et al. 1991) at 30 °C for various times and the reaction products were monitored using both HPAEC and high-performance size-exclusion chromatography.

3 HPAEC in Oligosaccharide Analysis

The HPAEC analysis of sugar oligomers is based on anion exchange of sugar anions generated at high pH, typically pH 13. The solvent system used in carbohydrate analysis generally consists of different gradients of 0.1 mol/l aqueous sodium hydroxide and 1 mol/l sodium acetate in 0.1 mol/l aqueous sodium hydroxide. Examples of typical gradients are given in Table 1. Two different column materials are indicated, i.e. CarboPac PA1 and PA100. The latter material provides the elution of sugar oligomers at somewhat lower sodium-acetate concentrations because of a lower capacity. From the PA1 column, the α-1,4-glucose decamer elutes at a sodium concentration of 0.38 mol/l and from the PA100 column at 0.35 mol/l, using the same gradient program.

As explained below, the total sodium concentration at which an oligosaccharide elutes is of utmost importance in HPAEC-MS coupling. The increasing sodium acetate concentration during the gradient elution is thought to act primarily as a pusher or displacer. The retention times or capacity ratios of oligomers in a homologous series obtained under two different gradient programs cannot be directly interrelated via the sodium concentration. In principle, a particular oligomer elutes at a relatively higher sodium-acetate concentration in a steep gradient program than in a flat program, but the steeper program results in a shorter analysis time. For example, the β-1,4-galactose pentamer elutes at a sodium concentration of 0.16 mol/l in the "arabinogalactan-1" program (see Table 1) after 35 min and at 0.32 mol/l in the steeper "arabinogalactan-2" program after 18 min.

The various oligomers generally elute in the order of increasing degree of polymerization (DP). However, large differences may exist in the elution time within a given gradient program when various series of oligomers are compared. A plot of the sodium concentration in the mobile phase at which a particular oligomer of a homologous series elutes from a CarboPac PA100 column in a standard gradient program (see Table 1) as a function of DP is shown in Fig. 2 for

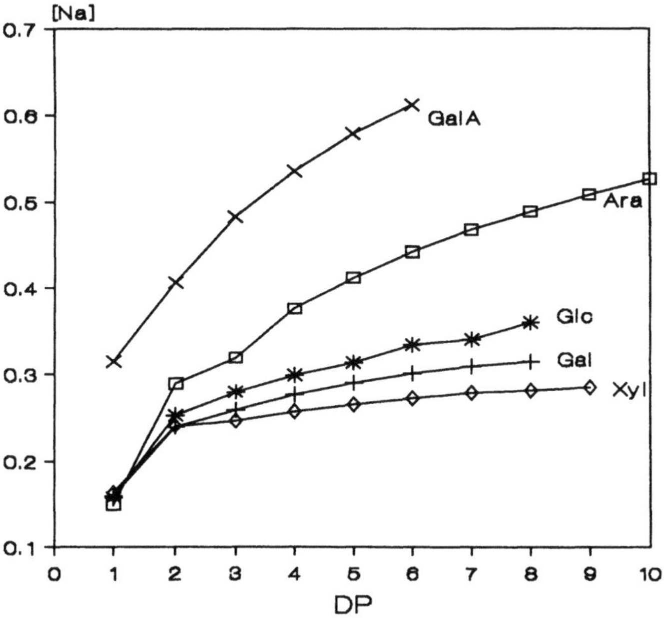

Fig. 2. Total sodium concentration [Na] at which sugar oligomers elute in HPAEC as a function of the degree of polymerization (DP). Standard gradient (cf. Table 1). CarboPac PA100 column. Homologous series of *Xyl* = β-1,4-xylose; *Gal* = β-1,4-galactose; *Glc* = α-1,4-glucose; *Ara* = α-1,5-arabinose; *GalA* = galacturonic acid

β-1,4-xyloses, α-1,5-arabinoses, β-1,4-galactoses, α-1,4-glucoses, and α-1,4-galacturonic acids. From this plot, it can be seen, for example, that the β-1,4-xylose pentamer elutes at a sodium concentration of 0.25 mol/l, while the galacturonic acid pentamer elutes at 0.58 mol/l. In practice, optimized problem-oriented gradient programs are developed, examples of which are given in Table 1.

For mixed oligomers, consisting of various sugar residues in either a linear or a branched conformation, it becomes more difficult or even impossible to predict the elution order of the various oligomers. As shown below, HPAEC-MS rapidly elucidates the elution orders of components in complex mixtures without the need for standards. It even provides information, for instance, on co-eluting sample constituents that are not readily obtained by other means.

4 General Experimental Considerations Related to HPAEC-MS

With thorough experience in thermospray mass spectrometry, the coupling of HPAEC and MS is relatively straightforward. Nevertheless, attention should be paid to several important aspects, e.g. on-line desalting of the HPAEC eluents and ionization of the oligosaccharides in thermospray mass spectrometry. The practical aspects discussed here are summarized from a number of papers on LC-MS of oligosaccharides from the authors' laboratories (Niessen and van der Greef 1992; Niessen et al. 1992a–d; van der Hoeven et al. 1992; Schols et al. 1994). Some aspects are dealt with in more detail in the original papers.

4.1 Desalting by AMMS

Gradients applied in oligosaccharide separation by HPAEC often require up to 0.5 mol/l sodium acetate (cf. Table 1). This results in mobile phases containing up to 0.6 mol/l of sodium. Sodium salts are involatile and their routine use in mobile phases for LC-MS, even at concentrations as low as 10^{-3} mol/l, is considered to be impossible because of rapid ion source contamination. In ion chromatography, similar problems occur since sodium carbonate gradients result in enormous baseline drifts in conductometric detection. An (anion) micromembrane suppressor (AMMS) system was introduced to solve these problems (Stillian 1985; Lee 1990). The function of the suppressor is to stoichiometrically exchange sodium ions in the column effluent for regenerant hydronium ions, thus effecting a neutralization reaction of the acidic counterion. The ion exchange is effected at a strong acid ion-exchange membrane, which is continuously regenerated with a sulphuric acid solution flowing at the opposite side of the membrane. In the gradients used in oligosaccharide analysis, the suppression results in the formation of acetic acid in the mobile-phase stream.

Being a membrane system, the AMMS has a back-pressure limitation of ca. 0.7 MPa. Since the operation of the thermospray requires a solvent pressure of ca.

4 MPa, repressurization of the effluent from the AMMS prior to introduction in the thermospray interface is required. This is done by means of a so-called booster pump (Simpson et al. 1990; cf. Fig. 1).

Being an ion-exchange system, the AMMS has a limited capacity for the sodium ions to be exchanged. Residual sodium concentrations after a single-stage AMMS were found to vary between 0.01 and 0.1 mmol/l with gradients of 0–0.2 mol/l of sodium acetate in 0.1 mol/l sodium hydroxide. In practice, using two AMMS in series, it was found that up to 0.4 mol/l of sodium ions could be effectively removed for LC-MS applications. This limits the gradient programs to be applied to a maximum of 30% solvent B. As a result, oligosaccharides needing higher concentrations of sodium acetate to elute are excluded from HPAEC-MS analysis. This explains the attention paid to the sodium-acetate concentrations in the previous section.

For a particular homologous series of sugar oligomers, the highest DP value that can be successfully analyzed in HPAEC-MS depends primarily on the sodium concentration at which that oligomer elutes and the concentration of that oligomer in the sample. For some homologous series, the highest DP value achievable has been explored. The results are summarized in Table 2. Mass chromatograms for the β-1,4-xylose-oligomers with DP = 23 – 25 are shown in Fig. 3. The DP = 25 xylose oligomer corresponds to a molecular mass of 3318; doubly charged disodium ions are detected (see below).

4.2 Ionization of Oligosaccharides

The thermospray interface, used to couple HPAEC and MS, is commonly operated with aqueous/organic solvent mixtures containing 10–100 mmol/l ammonium acetate. The use of ammonium-containing mobile phases in the thermospray MS analysis of oligosaccharides, in our experience, results in extensive breakdown to monomers, as a result of thermally induced ammoniolysis (Niessen et al. 1992b). Detection of intact oligosaccharides in thermospray MS appears to be possible by the addition of low concentrations of alkali acetates to the aqueous mobile phase, typically below 1 mmol/l. In the on-line combination of HPAEC-MS, the use of

Table 2. Maximum DP values observed in HPAEC-MS for oligosaccharides in homologous series

Sugar type	DP_{max} CarboPac PA1 column	DP_{max} CarboPac PA100 column
α-1,4-Glucose	6	14
α-1,5-Arabinose	9	–
β-1,4-Xylose	25	–
β-1,4-Galactose	5	12
Galacturonic acid	3	–

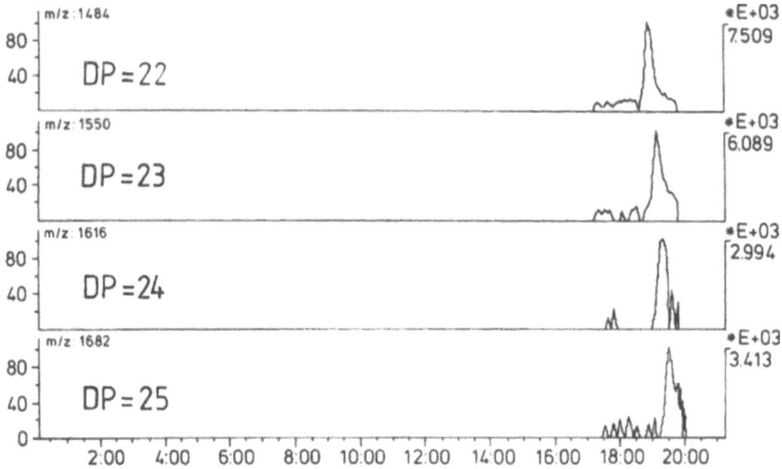

Fig. 3. HPAEC-MS of β-1,4-xylose oligomers. Mass chromatogram of doubly charged disodium ions for DP = 22–25, obtained by multiple-ion detection. CarboPac PA1 column; gradient, see Table 1

sodium acetate for this purpose is most obvious. When sodium-acetate concentrations in excess of 1 mmol/l are used, rapid ion source contamination within minutes prevents successful measurements.

The positive-ion mass spectra of neutral oligosaccharides obtained under these conditions contain strong peaks due to sodiated molecules, i.e. $[M+Na]^+$ at $m/z = M_r + 23$ and a weaker peak at $m/z = M_r + 45$ due to the $[M-H+2Na]^+$ ion, while for larger oligosaccharides (DP > 4) doubly charged disodium ions are observed, i.e. $[M+2Na]^+$ at $m/z = (M_r + 46)/2$, as well. Triply charged ions are not observed. These ionization conditions do not result in fragmentation (Niessen et al. 1992b,c).

The positive-ion mass spectra of acidic oligosaccharides, i.e. uronic acids, under the conditions generally used contain peaks due to the sodiated acid, $[M+Na]^+$ at $m/z = M_r + 23$, the sodiated sodium salts, e.g. $[M-H+2Na]^+$ at $m/z = M_r + 45$ and a sodiated fragment peak due to the loss of water from the sodiated acid, $[M+Na-H_2O]^+$ at $m/z = M_r + 5$. For acidic oligosaccharides containing more than one uronic acid, a corresponding number of additional peaks due to sodiated sodium salts are observed, e.g. $[M-2H+3Na]^+$ and $[M-3H+4Na]^+$ for an uronic acid trimer. From these additional peaks the number of acidic sugar residues in an oligosaccharide can be determined (Niessen et al. 1992c).

The expected m/z ratios for particular sodiated sugar oligomers can be calculated using the values in Table 3. The sugar monomers given in Table 3 are the prime sugars present in the type of samples analyzed thus far. The calculations are based on the m/z ratio for various monomeric sugars. For higher DP values, the m/z ratio is readily calculated using the sugar residue masses in Table 3. For example, an oligomer consisting of $Glc_4Xyl_2GalFuc$ is expected to give peaks at $m/z = 203 + 3 \times 162 + 2 \times 132 + 1 \times 162 + 1 \times 146 = 1261$ due to the sodiated

Talbe 3. m/z Values of sodiated oligosaccharides

Sugar type	Example	m/z of monomer	Residue mass (1+)	Residue mass (2+)
Aldopentose	Xylose (Xyl) Arabinose (Ara)	173	132	66
Deoxyaldohexose	Rhamnose (Rha) Fucose (Fuc)	187	146	73
Aldohexose	Glucose (Glc) Galactose (Gal) Mannose (Man)	203	162	81
Ketohexose	Fructose (Frc)	203	162	81
Hexuronic acid	Glucuronic acid (GlcA) Galacturonic acid (GalA)	217 239	176	88
4-O-methyl-hexuronic acid	4-O-methyl glucuronic acid (4-OMe-GlcA)	231 253	190	95

molecule, and at m/z = (1261 + 23)/2 = 642 for the doubly charged disodiated molecule. A small computer program has been written to either calculate molecular masses and expected m/z ratios of oligosaccharides with known sugar composition or to predict possible sugar compositions from measured m/z ratios.

Some preliminary experiments have also been performed with negative-ion thermospray MS of neutral oligosaccharides under HPAEC-MS conditions. The negative-ion mass spectra are surprisingly complex due to the formation of a variety of adduct ions. The most intense ions in the mass spectra of oligomers with DP < 5 are due to $[M-H]^-$, $[M+OAc]^-$ and $[M+HSO_4]^-$, while for larger oligomers in addition a variety of doubly charged ions are detected as well, with $[M+SO_4]^{2-}$ being the most prominent (Niessen et al. 1992d). The use of the negative-ion mode is still in development. It has not yet been used in the analysis of samples derived from plant cell walls. However, since the negative-ion mode provides a considerably better response for α-1,4-glucose oligomers than the positive-ion mode, especially at larger DP values, more extensive application of the negative-ion mode in this research area can be anticipated.

4.3 Data Interpretation

A significant background signal is observed in HPAEC-MS, especially during gradient elution at sodium concentrations exceeding 0.25 mol/l. This signal is due to the formation of sodiated sodium acetate cluster ions, leading to peaks at m/z = 187, 269 and subsequent m/z values at a 82-dalton increment. As a result, generally no peaks are observed in the total ion-current chromatograms. Actual analyte peaks must be searched for using reconstructed mass chromatography at calculated m/z values of the oligosaccharides expected to be present in the sample. This procedure is relatively easy as the number of possible m/z values is rather limited. For completely unknown oligosaccharides, this procedure of course is rather tedious and time-consuming. Often clean mass spectra also cannot be obtained. However, since no fragmentation is observed, the mass spectra do not contain additional information.

In practice, most samples are run twice. The first analysis is performed under full-scan acquisition conditions. After searching through the data using mass chromatography to detect some of the oligomers present, the sample is re-run under multiple-ion detection (MID) conditions using 5–10 ions per MID segment and switching from one segment to another at an appropriate moment during the elution. The latter is readily automated with the fully computer-controlled mass spectrometer used in these experiments. The MID procedure results in better signal-to-noise ratios and significantly improved peak shapes (Niessen et al. 1992d).

Since no fragmentation is observed, the HPAEC-MS technique in its present form primarily serves for the rapid molecular-mass determination of components in (complex) mixtures of sugar oligomers. At first, the determination of only molecular mass appears to be a serious limitation of the approach. For instance, no differentiation can be made between xylose and arabinose nor between glucose and galactose. However, the method has proved very useful in characterizing oligosaccharides in complex mixtures in a number of cases, e.g. in avoiding the need for complete fractionation, in helping in the interpretation of NMR spectra, and in identifying co-eluting peaks and elucidating the HPAEC elution order.

At present, routine analysis of unknown sugar oligomers with DP values up to 10–12 is possible. After determination of the molecular masses of the sample components, the sugar composition in terms of pentoses, deoxyhexoses, hexoses and (4-O-methyl) hexuronic acids and possibly other sugars can be determined using a small computer program. Perhaps surprisingly, the number of possible isomeric oligomers is rather limited, especially for oligomers containing none or only one acidic function as with most of the oligomers analyzed thus far. With the six types of monomers given in Table 2, a maximum of only seven isomers is found in determining sugar composition for m/z ratios corresponding to DP = 10 (none or only one acidic residue present). Combination of this information with knowledge on the origin of the sample and the type of enzyme used often leads to a unique sugar composition. In this respect, the data obtained by HPAEC-MS have been helpful in the interpretation of NMR spectra of fractionated compounds.

5 Application of HPAEC-MS in Oligosaccharide Characterization

The HPAEC-MS technology is developed in close contact with the application of the technique for the characterization of oligosaccharides enzymatically or chemically derived from plant cell wall polysaccharides. As a result, a variety of samples have been attacked, e.g. digests from apple, soy, maize and potato cell walls including rhamnogalacturonan, glucurono(arabino)xylan, xyloglucan and arabinogalactan oligomers. Most of the results with these types of samples are described in detail elsewhere (Niessen et al. 1992c,d; van der Hoeven et al. 1992; Schols et al. 1994a). In order not to repeat these results here, the discussion is focused on the results obtained in the analysis of arabinogalactans from potato and soy cell walls.

The dominant sugar residues present in the cell wall polysaccharides isolated from soybeans were arabinose, galactose, galacturonic acid and glucose. Cold sodium hydroxide was able to solubilize part of the pectic arabinogalactans present, concentrating all glucose as cellulose in the insoluble residue. Chromatography over DEAE-Sepharose CL6B demonstrated that in the sodium-hydroxide extract, various populations were present having an increased uronic acid content in the more strongly bound fractions, while the arabinose/galactose ratio varied. Methylation analysis revealed that the arabinose residues were present in highly branched arabinan chains. The galactose residues were mainly 1,4-linked and 25% of these residues were branched with both arabinose and galactose residues. Galacturonic acid was found to be present in highly ramified regions; homogalacturonan regions were hardly detected. Various fractions were incubated using pure galactanases and arabinanases in order to obtain more information on the distribution of substituents over the chain. Each enzyme released typical reaction products from their respective substrates and no synergism between the used enzymes was found. Incubation of the sodium-hydroxide extract by endo-galactanase over a short period of time resulted in a digest containing a homologous series of galactose oligomers, as concluded from the HPAEC analysis. Prolonged incubation resulted in monomeric, dimeric and trimeric end products, only some of which could be recognized on the basis of their retention behaviour on a CarboPac PA100 column. A HPAEC chromatogram of a typical digest, obtained after treatment of the arabinogalactans by endo-galactanase for 40 min, is shown in Fig. 4. In the HPAEC chromatogram of this digest, recorded using the PAD, a series of minor peaks are detected in between the peaks of the major components, the β-1,4-galactose oligomers. HPAEC-MS was used to identify these peaks. Using full-scan acquisition, the β-1,4-galactose oligomers in a particular sample could be detected up to DP = 7. Singly charged sodiated molecules were detected over the whole range, while for the DP = 6 and 7 doubly charged disodiated molecules were also detected. Between these peaks a second series of oligomers was detected at m/z values 30 dalton (for singly charged ions) lower than those of the galactose oligomers, thus indicating the presence of a galactose oligomer branched with a pentose, most likely an arabinose. The sample was also screened for other possible oligomers, for example, oligomers containing two pentose units, an uronic acid, or both a pentose and an uronic acid. None of these were detected.

Unfortunately, the full-scan acquisition resulted in poor peak shapes and did not unambiguously indicate whether various isomers of a particular Gal_nAra_1 were

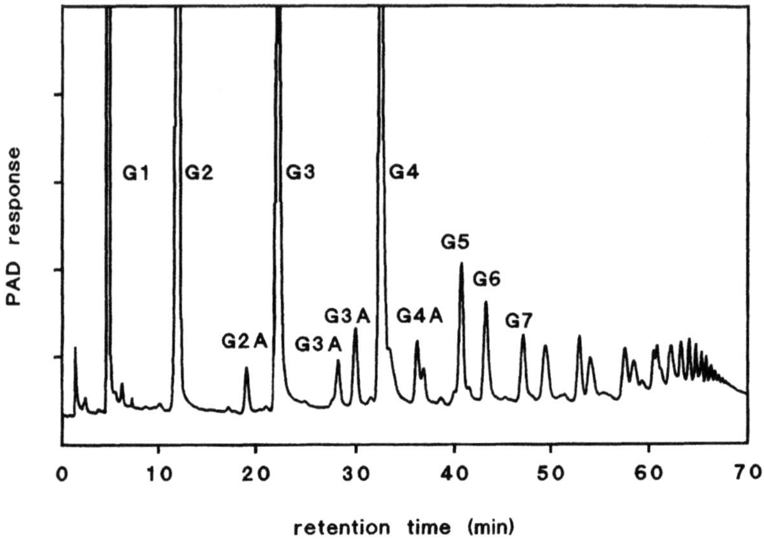

Fig. 4. HPAEC of an arabinogalactan digest from soybean cell wall material. CarboPac PA100 column, arabinogalactan-1 gradient (cf. Table 1)

Table 4. Number of galactose and arabinogalactose oligomers detected

n	Gal_n	Gal_nAra_1	Gal_nAra_2
0	–	1	0
1	1	2	1
2	2	3	2
3	2	2	0
4	1	1	2
5	1	1	2
6	1	1	3
7	1	1	1
8	1	1	1
9	1	1	1
10	1	1	1
11	1	1	
12	1		

present. As indicated above, both peak shape and sensitivity can be significantly improved by using multiple-ion detection (MID) instead of full-scan acquisition (Niessen et al. 1992d). With multiple-ion detection, it was possible to determine the number of isomeric oligomers present, at least as long as chromatographic separation of the isomeric oligomers is achieved. The results are summarized in Table 4. In comparison to the full-scan acquisition, the MID analysis also reveals the presence of minor constituents containing two arabinose units, i.e. Gal_nAra_2. At the higher DP values, the HPAEC using the gradient conditions indicated in Table 1 is no longer capable of resolving the various isomeric species.

From the peak shape of the Gal_2-oligomer it was expected that possibly two peaks were present. After dilution of the sample and re-injection, indeed two components are detected, which are found in a ca. 1:15 peak area ratio. The presence of two galactose dimers reveals the structural complexity of the original polysaccharide. While one of the two dimers originates from the linear β-1,4-linked galactan backbone, the minor component, which must be 1,2-, 1,3- or 1,6-linked, originates from a branched backbone residue. This observation is confirmed by permethylation analysis.

A series of experiments was performed on arabinogalactan samples which only had a short treatment with the β-1,4-galactanase enzyme. As expected, the oligomers with higher DP values are found to be present in more favourable concentrations in these samples. Various MID segments have been used to characterize this arabinogalactan digest. Special attention was paid to the oligomers with DP in the range of 6–12. A number of interesting observations have been made.

In the samples, oligomers like Gal_n, Gal_nAra_1 and Gal_nAra_2 were readily detected over the complete DP range covered. However, no Gal_nAra_3 oligomers have been detected. This indicates that such oligomers are either present in very low amounts or completely absent. Such indications can be translated towards polysaccharide structure in terms of the average number of arabinose side chains per number of galactose oligomers and to determine the mode of action of the enzyme used.

Despite the high resolution in HPAEC for various oligomers having the same DP value, the presence of other oligomers co-eluting with the β-1,4-galactose oligomers cannot be ruled out under the presently used gradient conditions and PAD. Using the combination of HPAEC-MS, the co-elution of compounds differing in molecular mass is readily detected. This is a clear example where the combination of HPAEC-MS rapidly provides information which is not readily obtained by any other means.

As indicated in Fig. 2, a homologous series of β-1,4-galactose oligomers elutes at lower sodium concentrations than an α-1,5-arabinose series. In Fig. 5, the retention time data for galactose and galactose-arabinose oligomers under two different gradient conditions are plotted as a function of DP. From these data it can be concluded that initially the presence of only one or two arabinose units significantly retains the mixed oligomer relative to the pure galactose oligomer. Thus, small oligomers, for instance Gal_2Ara and Gal_3, are readily separated with HPAEC under the present gradient conditions. This is not the case for similar oligomers with higher DP values, e.g. Gal_{10}, $Gal_{10}Ara$ and $Gal_{10}Ara_2$. The latter co-elute (cf. Fig. 5b). The presence of arabinose side chains significantly influences the retention time. The additional number of OH-groups introduced by the arabinose side chains counteract the increased retention due to the higher DP value. With the use of the MS, these co-eluting compounds can be readily resolved because of their difference in molecular mass. Typical mass chromatograms demonstrating these effects are given in Fig. 6.

This example not only clearly demonstrates the power of the HPAEC-MS to rapidly determine molecular masses of the sample constituents in multi-compo-

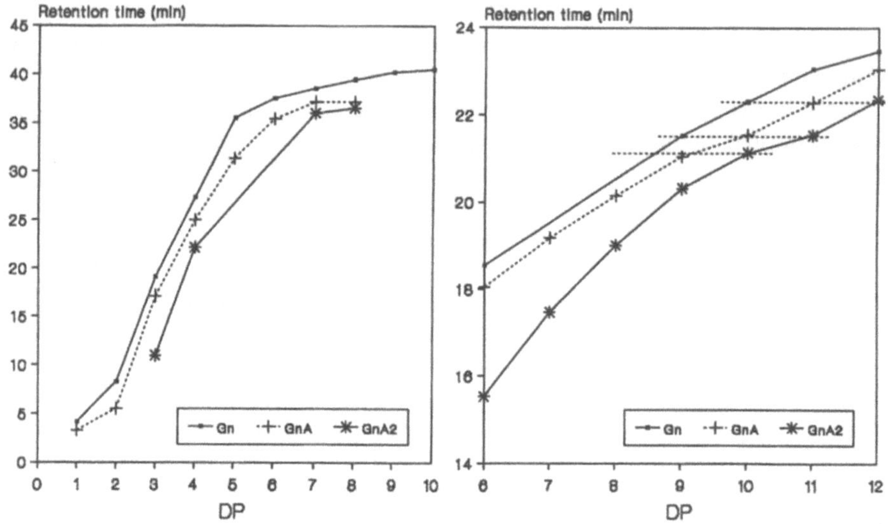

Fig. 5. Plot of retention times of Gal_n (*Gn*), Gal_nAra (*GnA*) and Gal_nAra_2 (*GnA₂*) oligomers from a arabinogalactan digest from soybean cell wall material as a function of DP. *Left* Arabinogalactan-1 gradient; *right* arabinogalactan-2 gradient (cf. Table 1). CarboPac PA100 column

nent mixtures, but also one of its weaknesses. Obviously, one would like to be able to determine the sugar residue sequence in the oligomers detected. The possibilities for using tandem mass spectrometry for this purpose are presently under investigation. Preliminary results indicate that this is feasible, although differentiation sometimes can only be made on the basis of differences in relative intensities between known isomers. Considerable effort should be invested in learning the fragmentation rules for various oligomers and their constituent sugar residues before this approach can be routinely used in carbohydrate characterization by HPAEC-MS. Further work in this direction is in progress.

6 Conclusions and Perspectives

In conclusion, the LC-MS analysis of oligosaccharides has proven to be a powerful technique. The HPAEC-MS approach provides rapid molecular-mass determination of the constituents of complex oligosaccharide mixtures obtained by enzymic degradation of plant cell wall polysaccharides up to a DP of 10–12. This is helpful in the identification of the degradation products of such processes. This information can be used either to determine the structure of the polysaccharide or to establish enzyme activity and specificity or both. Furthermore, the HPAEC-MS approach enables the detection of co-eluting species and the determination of the often unpredictable elution order of oligosaccharides in HPAEC.

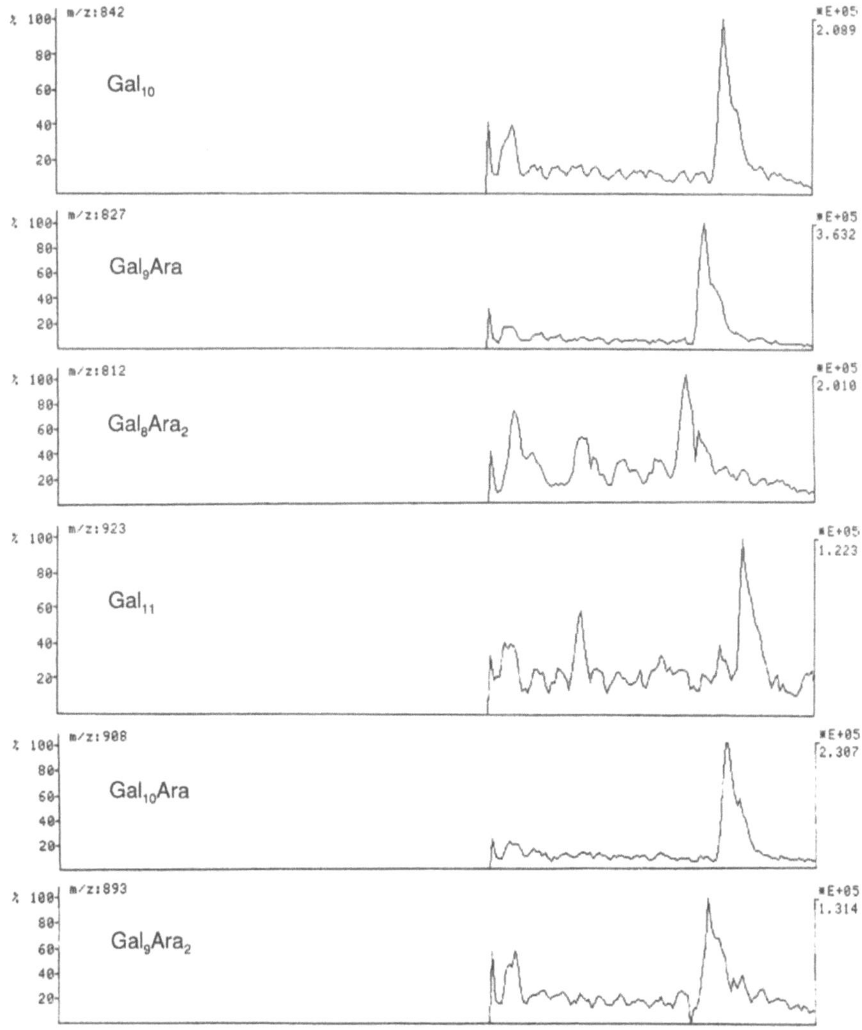

Fig. 6. HPAEC-MS of an arabinogalactan digest from soybean cell wall material. Mass chromatograms for doubly charged disodiated molecules of *Gal$_{10}$*, *Gal$_9$Ara*, *Gal$_8$Ara$_2$*, *Gal11*, *Gal$_{10}$Ara*, and *Gal$_9$Ara$_2$*. CarboPac Pa100 column; arabinogalactan-2 gradient (cf. Table 1)

During the execution of this project, a new and powerful LC-MS interface became commercially available, i.e. the so-called electrospray interface (Niessen and van der Greef 1992). Initially, this interface, which offers ionization characteristics similar to the thermospray interface, could only be used at flow rates in the low µl/min range. Therefore, coupling of HPAEC and MS via such a low flow-rate electrospray interface cannot be readily achieved for a variety of reasons. Anion-exchange microcolumns are not (yet) available, and redesigning of the AMMS is required in order to achieve considerably lower dead volumes than the present

40 μl. Furthermore, the electrospray process is even more sensitive to the presence of sodium in solution. However, in our laboratory, the electrospray interface has been used frequently and successfully in constant-infusion or flow-injection-analysis mode for the mass spectral characterization of fractionated sugar oligomers, e.g. prior to NMR analysis (Tinke et al. 1993). Recently, a new generation of electrospray interfaces has become available which allow the use of larger flow rates (up to 1 ml/min), thus enabling on-line HPAEC-MS coupling.

Another LC-MS interface strategy, somewhat similar to electrospray, is the ionspray approach. The on-line combination of HPAEC and MS using an ionspray interface for use in oligosaccharide characterization was described by the group of Henion (Conboy and Henion 1992). This research attempts to characterize oligosaccharides derived from glycoproteins.

More advanced LC-MS interface technology, such as electrospray and ionspray, are now available, which may allow even improved performance in HPAEC-MS coupling in the near future. Irrespective the LC-MS interface, various topics need further study in order to provide us with a more solid understanding of the ionization phenomena under the conditions applied in both the thermospray and the electrospray/ionspray interface. Such an understanding is expected to lead to improvements in the minimum detectable amounts, which presently are in the high picomole to low nanomole range. Another important topic for further research is the application of tandem mass spectrometry in conjunction with the present technology which may enable further elucidation of the structure of the oligosaccharides detected in HPAEC-MS, e.g. in terms of sugar residue sequence, linkage position and type of sugar residue.

Acknowledgements. The authors would like to thank C. Bruggink (Dionex) for helpful discussions and his support and Dionex B.V. (Breda, The Netherlands) for supplying the HPAEC system used for this work. Part of this work was supported by the Netherlands Program for Innovation Oriented Cabohydrate Research (IOP-k) with financial aid from the Ministeries of Economic Affairs and Agriculture, Nature Management and Fisheries, Cehave NV (Veghel, NL), BP Nutrition Nederland (Boxmeer, NL) and Gist brocades (Delft, NL).

References

Arpino PJ (1990) Combined liquid chromatography mass spectrometry. Part II. Techniques and mechanism of thermospray. Mass Spectrom Rev 9:631–669

Conboy JJ, Henion JD (1992) High-performance anion-exchange chromatography coupled with mass spectrometry for the determination of carbohydrates. Biol Mass Spectrom 21:397–407

Düsterhöft EM, Voragen AGJ, Engels FM (1991) Non-starch polysaccharides from sunflower (*Helianthus annuus*) meal and palm kernel (*Elaeis guineenis*) meal – preparation of cell wall material and extraction of polysaccharide fractions. J Sci Food Agric 55:411–422

Gruppen H, Marseille JP, Voragen AGJ, Hamer RJ, Pilnik W (1989) Mild isolation of water-unextractable cell wall material from wheat flour. Composition of fractions obtained with emphasis on non-starch polysaccharides. J Cereal Sci 9:247–260

Gruppen H, Hoffmann RA, Kormelink FJM, Voragen AGJ, Kamerling JP, Vliegenthart JFG (1992) Characterization by ^{1}H NMR spectroscopy of enzymically derived oligosaccharides from alkali-extractable wheat-flour arabinoxylan. Carbohydr Res 233:45–65

Lee YC (1990) High-performance anion-exchange chromatography for carbohydrate analysis. Anal Biochem 189:151–162

Niessen WMA, van der Greef J (1992) Liquid chromatography mass spectrometry. Marcel Dekker, New York

Niessen WMA, van der Hoeven RAM, van der Greef J (1992a) Analysis of intact oligosaccharides by liquid chromatography mass spectrometry. Org Mass Spectrom 27:341–342

Niessen WMA, van der Hoeven RAM, van der Greef J, Schols HA, Voragen AGJ (1992b) On-line liquid chromatography thermospray mass spectrometry in the analysis of oligosaccharides. Rapid Commun Mass Spectrom 6:197–202

Niessen WMA, van der Hoeven RAM, van der Greef J, Schols HA, Lucas-Lokhorst G, Voragen AGJ, Bruggink C (1992c) High-performance anion-exchange chromatography thermospray mass spectrometry in the analysis of oligosaccharides. Rapid Commun Mass Spectrom 6:474–478.

Niessen WMA, van der Hoeven RAM, van der Greef J, Schols HA, Voragen AGJ, Bruggink C (1992d) Recent progress in high-performance anion-exchange chromatography thermospray mass spectrometry of oligosaccharides. J Chromatogr 647:319–327

Rajakylä E (1986) Use of reversed-phase chromatography in carbohydrate analysis. J Chromatogr 353:1–12

Reinhold VN (1986) Structural elucidation of complex carbohydrates. In: Gaskell SJ (ed) Mass spectrometry in biomedical research. Wiley, London, pp 181–213

Santikarn S, Her GR, Reinhold VN (1987) Oligosaccharide structural studies by on-line HPLC-MS using fast atom bombardment ionization. J Carbohydr Chem 6:141–154

Schols HA, Mutter M, Voragen AGJ, Niessen WMA, van der Hoeven RAM, van der Greef J, Bruggink C (1994a) The use of combined high-performance anion-exchange chromatography thermospray mass spectrometry in the structural analysis of pectic oligosaccharides. Carbohydn Res 267:335–342

Schols HA, Voragen AGJ, Colquhoun IJ (1994b) Isolation and characterization of rhamnogalacturonan oligomers, liberated during degradation of pectic hairy regions by rhamnogalacturonase. Carbohydr Res 256:97–111

Selvendran RR, Ryden P (1990) Isolation and analysis of plant cell walls. In: Dey PM (ed) Methods in Plant Biochemistry Academic Press, London 2:549–579

Simpson RC, Fenselau CC, Hardy MR, Townsend RR, Lee YC, Cotter RJ (1990) Adaptation of a thermospray liquid chromatography/mass spectrometry interface for use with alkaline anion exchange liquid chromatography of carbohydrates. Anal Chem 62:248–252

Stillian J (1985) An improved suppressor for ion chromatography. LC Mag 3:802–805

Tinke AP, van der Hoeven RAM, Niessen WMA, van der Greef J, Vincken J-P, Schols HA (1993) Electrospray mass spectrometry of neutral and acidic oligosaccharides: methylated cyclodextrins and identification of unknowns derived from fruit material. J Chromatogr 647:279–287

Van der Hoeven RAM, Niessen WMA, Schols HA, Bruggink C, Voragen AGJ, van der Greef J (1992) Characterization of sugar oligomers by on-line high-performance anion-exchange chromatography thermospray mass spectrometry. J Chromatogr 627:63–73

Van de Vis H, Searle-van Leeuwen MJF, Siliha HA, Kormelink FJM, Voragen AGJ (1991) Purification and characterization of endo-1,4-β-D-galactanase from *Aspergillus niger* and *A. aculeatus*: use in combination with arabinases from *A. niger* in enzymic conversion of potato arabinogalactan. Carbohydr Polymers 16:167–187

Verbruggen MA, Beldman G, Voragen AGJ, Holleman M (1993) Water-unextractable cell wall material from sorghum: isolation and characterization. J Cereal Sci 17:71–82

Voragen AGJ, Schols HA, Gruppen H (1993) Structural studies of plant cell wall polysaccharides using enzymes. In: Meuser F, Manners DJ, Seibel W (eds) Plant polymeric carbohydrates. R Soc Chemistry, Cambridge, pp 3–13

Analysis of Pectin Methyl Esterases

MARIANNE BORDENAVE

1 Introduction

Pectins represent one of the major components of primary cell walls of dicot plants. They consist of a (1,4)-linked α-D-galacturonan chain backbone interspersed with (1,2)-linked α-L-rhamnose units that cause bends in the pectic molecule (Fig. 1); a proportion of the galacturonic acid units can be methoxylated at C_6. Rhamnose-rich segments are often branched with arabinogalactan side chains, and usually exhibit a high degree of methylation. Unbranched blocks, with few rhamnose units, alternate with these highly branched blocks (Jarvis 1984). Pectins are synthesized inside the cell, in the Golgi apparatus (Zhang and Staehelin 1992). First, the backbone is synthesized, then the methyl groups are transferred from S-adenosyl methionine to the carboxyl groups giving rise to highly methylated galacturonans which are then exported to the cell wall. It is generally suggested that plant pectins undergo demethylation when they are inside the wall. Pectin methylesterase (PME) [EC 3.1.1.11], one of the cell wall enzymes, is thought to catalyze this demethylation reaction (Fig. 2).

PME is a very ubiquitous enzyme, found in plants, fungi and bacteria (Rexova-Benkova and Markovic 1976; Versteeg 1979); also, the presence of PME activity in salivary secretions of greenbugs has been recently reported (Ma et al. 1990). In microorganisms, PMEs are generally secreted outside the cell during phytopathogenic processes, together with other cell-wall degrading enzymes, like cellulases, polygalacturonases, pectin and pectate lyases, and proteases (Kotoujanski 1987).

Because of their role in the gelification of plant pectins, in fruit maturation and possibly in phytopathogenic processes, PMEs have been widely studied, and investigations on this enzyme are still in progress.

2 Estimation and Detection of PME Activities

2.1 Detection

PME can be rapidly detected in solution by incubating an aliquot of the enzyme solution in a mixture containing pectin and a pH indicator. When PME is present

Modern Methods of Plant Analysis, Vol. 17
Plant Cell Wall Analysis
Edited by H.-F. Linskens and J.F. Jackson

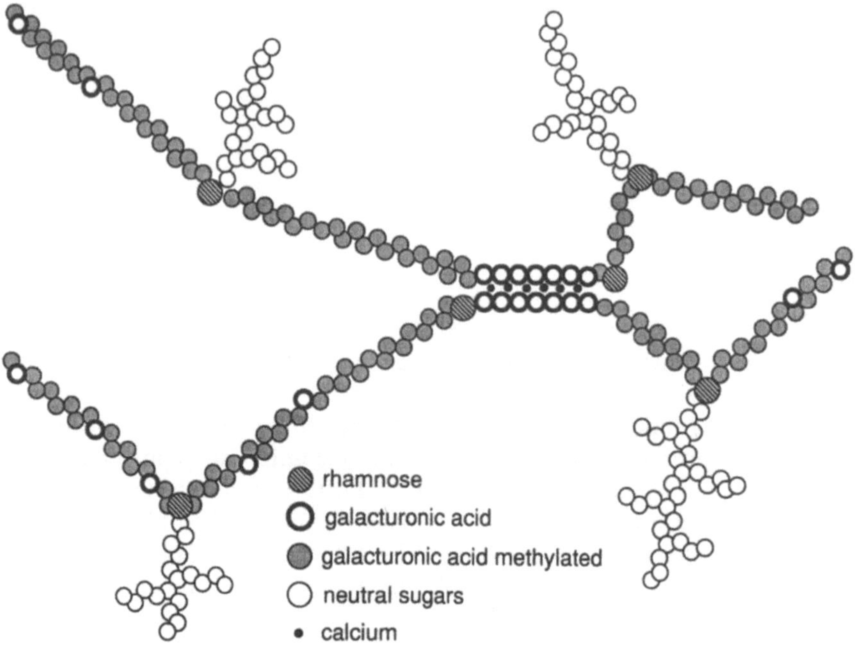

Fig. 1. Structure of pectins. (Courtesy of Dr. Roger Prat)

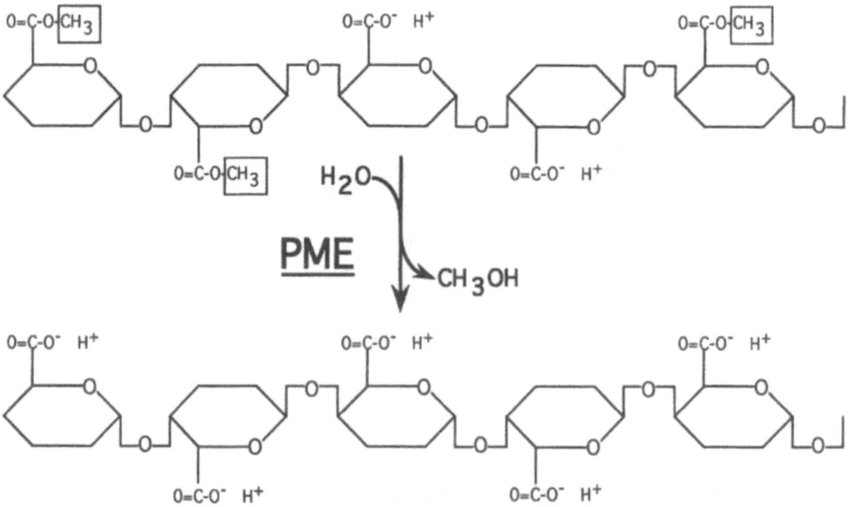

Fig. 2. Pectin methylesterase reaction

in the aliquot and hydrolyzes pectin methyl groups, it produces protons, thus inducing a colour change in the solution. Methyl red as pH indicator is very convenient for this kind of detection (Bordenave et al. 1993).

PME activity can also be easily detected after non-denaturing electrophoresis like IEF (iso-electrofocalization) or native PAGE (polyacrylamide gel electrophoresis). These methods allow the detection of PME molecules among other proteins and also, in the case of IEF, the determination of their pI. The PME-containing gel is applied on a substrate gel, often an agar pectin gel. After incubation (generally at 37 °C), both parts of this "sandwich" are incubated in ruthenium red, a dye specific for acidic pectins (Bertheau et al. 1984). Bands corresponding to PME isozymes appear in the substrate gel as dark-red spots on a lighter background, and as pink spots in the migration gel. Methylene blue has also been used in a similar manner (Roeb and Stegemann 1975).

2.2 Estimation

PMEs, which are mostly ionically bound to the cell walls (Giovane et al. 1990), can be easily removed from their natural matrix by incubating cell walls in saline solutions of high ionic strength. Both bound and solubilized PME activities can be estimated by automatic titration of the H^+ produced by the reaction, in a solution containing the enzyme and its substrate (Lee and MacMillan 1968). Free PME activity can also be estimated spectrophotometrically by monitoring the colour changes of an indicator dye added to the reaction mixture (Hagerman and Austin 1986). These methods allow easy measurement of the enzyme reaction rate, but in the second case, the pH of incubation is not constant and, furthermore, the measures can only be performed at a pH corresponding to the colour change of the dye.

An alternative method has been proposed for measuring enzyme activity in solution. It consists in monitoring the de-esterification of *para*-nitrophenylacetate, a substrate for all esterases (Zimmerman 1978). This method is not very specific, but allows estimation of PME activities even in buffered solution, which can be interesting when dialysis cannot be performed. Nevertheless, it is now evident that some PMEs do not de-esterify *p*NPA (*para*-nitrophenylacetate), and that some extracts very active on *p*NPA do not contain any PME (Bordenave and Goldberg 1993).

Also, methods for measuring methanol produced by PME activity have been developed, but they are not as easy to perform (Woods and Siddiqui 1971). Lastly, a method for a pH-independent assay of PME has been proposed by MacMillan et al. (1993), but it appears slightly complex, involving eight different steps before the actual determination of PME activity, which renders the estimation of the initial reaction rate very difficult.

3 PME Localization

In plant cells, PME activity appears to be associated with the cell wall. When plant cells are ground and cellular components fractionated, most of the PME activity is found in the cell wall fraction, and very little in the microsomal fraction. In this case, the PME activity probably arises from enzyme molecules extracted in the course of their transportation process through the Golgi apparatus (Goldberg 1984). Because of their apoplastic localization and their modifying wall structure activity, PMEs have often been studied with regard to their relations with plant development processes, like elongation of plant organs (Moustacas et al. 1986; Nari et al. 1991; Gaffé et al. 1992; Bordenave and Goldberg 1993) or maturation of fruits (Ray et al. 1988; Harriman et al. 1991; Tieman et al. 1992).

Few immunocytolocalization experiments have been performed with antisera raised against PME molecules (see e.g. Goldberg et al. 1992a). Such experiments, however, indicate that PMEs are mainly present in the middle lamella and in the cell junctions, and cytochemical methods have shown that these parts of the plant cell wall are very rich in acidic pectins (Goldberg et al. 1986). In contrast, methylated pectins appear to be distributed throughout the cell walls.

The differential localization of PMEs and their substrates might explain why plant cell walls can contain high amounts of methylated pectins, together with enough PME molecules theoretically to demethylate these pectins entirely. In fact, most PMEs are tightly bound to cell walls, and thus are probably unable to demethylate the highly methylated pectins that somehow stay beyond their reach, unless a favourable ionic environment allows them to move in other parts of the wall. In mung bean hypocotyl, in which several PME isoforms were detected and purified, it was observed that the fixation strength varied greatly from one isoform to another. The weakly bound isoforms might be the most active in vivo, because it is difficult to imagine how enzymes strongly linked to the cell wall matrix are able to de-esterify cell wall pectins (for example, in *Vigna radiata* hypocotyls, one of the major PME isoforms requires 1 M NaCl to be solubilized). In conclusion, plants might regulate PME activity through control of the ionic concentrations inside the cell wall. One can imagine that inside the cell wall, micro-domains where ionic concentrations are adequate for enzyme liberation and reaction, might co-exist with domains where PME is retained and cannot function.

4 PME Extraction and Purification

PMEs are usually solubilized from plant cell walls by saline treatment. First, the organ(s) of interest is sectioned and ground in a buffer solution; cellular components are then removed by several washings, and by treatment with a mild detergent (e.g. Triton X-100, 0.1%). The cell wall suspension is then submitted to a high salt concentration (usually 1 M NaCl), so that proteins interacting ionically with the matrix are solubilized. In *Vigna radiata*, different PME isoforms have been eluted sequentially by treating cell walls with NaCl solutions of increasing concentration (Bordenave and Goldberg 1994). After saline treatment, cell walls are removed by filtration. The saline eluate can then be fractionated by ammonium sulphate precipitation and/or various chromatographic methods (often anion or cation exchange, at least for the first steps of purification) (Hultin et al. 1966; Brady

1976; Lourenco and Catutani 1984; Markovic et al. 1988; Lin et al. 1989; Komae et al. 1990; Proctor and Miesle 1991; Gaffé et al. 1992; Pressey and Woods 1992; Bordenave and Goldberg 1993; Lim and Chung 1993).

5 PME Properties

5.1 Physicochemical Properties

The physicochemical properties of already purified plant or microorganism PMEs are summarized in Tables 1 and 2.

The existence of multiple PME isoforms has been reported in orange (Versteeg et al. 1978), tomato (Pressey and Woods 1992), kiwi (Giovane et al. 1990), grapefruit (Seymour et al. 1987), banana (Brady 1976), papaya (Lim and Chung 1993), lemon (Evans and MacHale 1978), flax (Gaffé et al. 1992) and mung bean (Bordenave and Goldberg 1993). Usually, these isoforms differ in their isoelectric point and/or their molecular mass and catalytic properties. Relative proportions of these isoforms may vary greatly according to the development stage and the organ considered; e.g. in *Vigna radiata*, the relative quantities of the four major PME isoforms vary along the hypocotyl (Bordenave and Goldberg 1993). Also, PME isoforms can be specific for one organ and/or a precise development stage; e.g. a PME from *Brassica napus* is only expressed in the developing pollen (Albani et al. 1991), and in tomato, one PME isoform is specifically expressed in the maturing fruit (Harriman et al. 1991).

Most plant PMEs present basic pIs (ca. 9). However, in *Ficus*, two acidic PMEs (pI 4.4 and 3.5) were purified respectively from red tepals (Komae et al. 1990) and achenes (Lin et al. 1989), and in *Vigna radiata* two acidic isoforms were detected in the lower part of the hypocotyl (Bordenave and Goldberg 1993). The molecular mass ranged between 21 000 and 53 000 daltons. Like many other cell wall proteins, PMEs can be glycosylated (Giovane et al. 1990; Seymour et al. 1991; Rillo et al. 1992); and in one case, PME was reported to be a lipoprotein (Lee and MacMillan 1968). It has been proposed, and in part demonstrated (Seymour et al. 1991), that these post-traductional modifications might increase PME stability.

In comparison with plants, microbial PMEs often present more acidic pIs, between 3.6 and 6.8 (Rexova-Benkova and Markovic 1976; Markovic et al. 1983; Sakellaris and Evangelopoulos 1989; Khanh et al. 1991), except for the phytopathogenic enterobacterium *Erwinia chrysanthemi*, the PME of which presents a pI of 9.64 (Laurent et al. 1993). Their M_r ranges from 27 800 to 48 000; they also differ from plant PMEs in their catalytic properties. Like plant PMEs, fungus PME can be glycosylated, and in one case a bacterial PME also appeared to be glycosylated (Van Rijssel et al. 1993). In some cases, PME activity is expressed by multi-enzymic complexes also exhibiting pectate lyase (PL) (Sheiman et al. 1976) or polygalacturonase (PG) (Van Rijssel et al. 1993; Schink and Zeikus 1983) activities.

Table 1. Physicochemical properties of purified plant PMEs

Species	Organ extracted	Molecular form(s)	M_r	pI	References
Medicago sativa	Leaves, stems, and flowers	5	34000	7.5 to 9.5	Markovic et al. (1988)
Malus malus	Fruit	1	28000	–	Versteeg (1979)
Musa acuminata	Fruit	2	30000	8.9	Brady (1976)
			30000	9.4	Brady (1976)
Ficus awkeotsang	Achenes	1	38000	3.5	Lin et al. (1989)
Ficus awkeotsang	Seeds	1	42000	4.4	Komae et al. (1990)
Linum usitatissimum	Callus	6	20000 to 43000	5.5 to >10	Gaffé et al. (1992)
Citrus paradisi	Fruit	2	36000	>10	Seymour et al. (1991)
			53500	>10	Seymour et al. (1991)
Actinidia chinensis	Fruit	2	57000	7.3	Giovane et al. (1990)
			57000	7.3	Giovane et al. (1990)
Citrus aurentifolia	Fruit	2	–	–	Evans and MacHale (1978)
Citrus reticulata	Fruit	1	37000	>8.65	Rillo et al. (1992)
Vigna radiata	Hypocotyl	4	53000	7.0	Bordenave and Goldberg (1993)
			53000	7.5	Bordenave and Goldberg (1993)
			40000	8.5	Bordenave and Goldberg (1993)
			40000	>10	Bordenave and Goldberg (1993)
Citrus sinensis	Fruit	1	36200	–	Versteeg (1979)
Citrus sinensis	Fruit	2	–	–	Evans and MacHale (1978)
Citrus sinensis	Fruit	3	36200	10.05	Versteeg et al. (1978)
			36200	>11	Versteeg et al. (1978)
			54000	10.2	Versteeg et al. (1978)
Carica papaya	Fruit	1	53000	–	Lourenco and Catutani (1984)
Carica papaya	Fruit	2	28000	>9	Lim and Chung (1993)
			27000	>9	Lim and Chung (1993)
Glycine max	Cell suspensions	1	33000	–	Moustacas et al. (1986)
Lycopersicon esculentum	Fruit	1	23000	–	Tucker et al. (1982)
Lycopersicon esculentum	Fruit	2	23800	9.3	Pressey and Woods (1992)
			24200	8.9	Pressey and Woods (1992)

Table 2. Properties of microorganism PMEs

Species	M_r	pI	K_m(g/l)	pH opt.	References
Fungi					
Aspergillus niger	43 000	3.6–3.8	–	4–4.5	Khahn et al. (1991)
Monilia laxa	–	–	–	4	Rexova-Benkova and Markovic (1976)
Coniothyrium diplodiella	–	–	–	4.5–5	Rexova-Benkova and Markovic (1976)
Fusarium oxysporum	–	–	–	5	Rexova-Benkova and Markovic (1976)
Sclerotinia libertiana	–	–	–	4.8–5.2	Rexova-Benkova and Markovic (1976)
Rhizoctonia solani	–	–	–	6–7	Rexova-Benkova and Markovic (1976)
Diplodia gossypina	–	–	–	6.5	Rexova-Benkova and Markovic (1976)
Cercosporella herpotrichoides	–	–	–	7.5	Rexova-Benkova and Markovic (1976)
Pellicularia filamentosa	–	–	–	7	Rexova-Benkova and Markovic (1976)
Acrocylindrium	–	–	–	7–7.5	Rexova-Benkova and Markovic (1976)
Phytophtora infestans	45 000	–	–	7	Forster (1988)
	48 000	–	–	7	Forster (1988)
Botrytis cinerea	28 400	6.8	0.006	7	Schejter and Marcus (1988)
	27 800	6.8	0.0004	6.5	Schejter and Marcus (1988)
Bacteria					
Pseudomonas prunicola	–	–	–	7.8	Rexova-Benkova and Markovic (1976)
Pseudomonas solanacearum	41 000	–	–	–	Spok et al. (1991)
Clostridium multifermentans	400 000	–	0.0735	9	Sheiman et al. (1976)
Clostrdium thermosaccharolyticum	320 000	–	1.2	–	Van Rijssel et al. (1993)
Clostridium thermosulfurogenes	320 000	–	–	6.5	Schink and Zeikus (1983)
Lactobacillus plantarum	35 000	5	2.1	7.2	Sakellaris and Evangelopoulos (1989)
	35 000	6	2.9	7.2	Sakellaris and Evangelopoulos (1989)

5.2 Enzymatic Properties

5.2.1 Action Pattern

PMEs are highly specific for methylesters of D-galacturonans, since they can act neither on pectins esterified by ethyl or propyl groups, nor on methyl alginates. De-esterification of pectins by PMEs is never complete; in oranges and tomatoes, the DE (degree of esterification) obtained after enzymic demethylation is about

11%, in alfalfa, it is about 3% (Rexova-Benkova and Markovic 1976; Versteeg 1979). It is very difficult to compare the kinetic parameters of PMEs, like K_m or V_{max}, found in the literature, since the estimations of these values are often made at different pHs, under different saline conditions, and with different substrates (Table 3).

As to the mode of action of PMEs, it appears that most plant PMEs de-esterify pectins blockwise. The de-esterification starts near free carboxyl functions or from the reducing end of the pectic chain (Solms and Deuel 1955; Rexova-Benkova and Markovic 1976; Markovic and Kohn 1984). In fact, because of the need of free carboxyl functions on the polygalacturonic chain, plant PMEs are much less active on a highly methylated pectin than on pectins presenting a DE of 20–30% (Evans and MacHale 1978; Seymour et al. 1991). In microbial PMEs, de-esterification can be developed blockwise, or randomly, depending on the species (Rexova-Benkova and Markovic 1976; Markovic and Kohn 1984). Consequences of these two types of de-esterification are very different: blockwise, it favours pectin gelification in the presence of calcium; randomly, it favours polygalacturonase and pectate lyase action and thus pectin depolymerization. Markovic and Kohn (1984) have proposed that the pI and pH optimum of the enzyme might be important in determining the mode of action of PME. It cannot be excuded that some plant PME isoforms might randomly de-esterify pectins. These isoforms may likely take part in processes where partial digestion of the cell wall is necessary, e.g. fruit ripening or cell wall elongation.

PME immobilization greatly enhances their activity towards exogenous substrate, and their solubilization often results in an important loss of activity. Consequently, it seems that fixation of PMEs on the cell wall induces conformational features important for their activity (Borrego et al. 1989).

Also, it has been proposed that in vivo, some PMEs might have a pectin methyl transferase activity, depending on the local environment (Glasziou 1959). So far, this problem has not been resolved, and will probably remain unsolved until the exact physicochemical conditions inside the cell wall have been determined.

5.2.2 Influence of pH on PME Activity

Plant PMEs appear to be very sensitive to their ionic environment, and are usually closely regulated by pH. Indeed, plant PMEs present pH optima between 6 and 8. This behaviour might be due to strong interactions, at acidic pHs, between the positively charged enzyme and the free carboxyl groups of polygalacturonic acid, a product of the enzymic reaction. At basic pHs, where plant PMEs are neutral or negatively charged, because of their basic pIs, it has been proposed that PME might be released from their substrate because of electrostatic repulsion between free carboxyl groups and the negatively charged enzyme (Nari et al. 1991).

Also, above pH 7, pectins undergo autode-esterification, which would in turn enhance PME activity, which needs free carboxyl groups to work. In contrast, reported microbial PMEs present pH optima between 4 and 9, and are often less sensitive to the ionic environment than plant PMEs.

Table 3. Enzymatic properties of some plant PMEs

Species	Substrate	Conditions	Specific activity (μKatal[a]/mg protein)	K_m (g/l)	V_{max} (μKatal/mg protein)	pH opt.	References
Ficus awkeotsang	Apple pectin	50 mM KCl, pH 7.5	–	1.19	1.96	7–8	Komae et al. (1990)
	Ficus pectin	50 mM KCl, pH 7.5	–	2.78	1.4	7–8	Komae et al. (1990)
	Apple pectin	No salt, pH 5.4	–	3.70	1.8	5.4	Komae et al. (1990)
	Ficus pectin	No salt, pH 5.4	1.2	2.94	1.1	5.4	Komae et al. (1990)
Ficus awkeotsang	*Citrus* pectin	0.2 M NaCl, pH 7, at 37 °C	4.8	0.75	5.17	6.5	Lin et al. (1989)
Linum usitatissimum	*Citrus* pectin	No salt, pH 7.7	3.3	0.147	7.1	6	Gaffé et al. (1992)
Actinidia chinensis	*Citrus* pectin	No salt, pH 7.5	15.5	1.82	–		Giovane et al. (1990)
			16.2	0.76	–		
Citrus aurantifolia	Pectin	0.1 M NaCl, pH 7, at 25 °C	2.4	0.015	–	9	Evans and MacHale (1978)
			5.5	0.054	–	6	
Citrus sinensis	Pectin	0.1 M NaCl, pH 7, at 25 °C	3.7	0.025	–	8	Evans and MacHale (1978)
			7.4	0.021	–	<6	
Carica papaya	Apple pectin (BDH)	0.06 M NaCl in 0.05 M sodium phosphate buffer, pH 7, at 35 °C	15.3	0.0071	12.30	8	Lim and Chung (1993)
			6.3	0.016	13.3	8	Lim and Chung (1993)
Carica papaya	*Citrus* pectin	0.2 M NaCl, pH 8, at 35 °C	0.6	0.12	–	8.0	Lourenco and Catutani (1984)
Lycopersicon esculentum	Pectin	0.1 M NaCl, pH 7, at 25 °C	14.5	–	–	7.5	Prssey and Woods (1992)
			14.8	–	–	7.5	

[a] 1 Katal = 1 mol substrate hydrolyzed per second.

PME solubilization can modify their pH optimum; for example, in *Vigna radiata* the pH optimum of PME activity of isolated cell walls was around 7.6, whereas PME activity of the saline cell wall extract did not show any obvious pH optimum between 5.6 and 7.6 (Bordenave and Goldberg 1993). Moreover, after purification of the four isoforms present in the saline extract, the two "neutral" isoforms (pI around 7.5) still presented no pH optimum, whereas the two cationic isoforms (pI 8.5 and pI > 9) presented respectively optima around pH 7.6 and 5.6, the latter being highly unusual, compared to the plant PME pH optimum most often reported in the literature (ca. 7.6).

5.2.3 Influence of Cations

In plant PMEs, cations appear to be essential for the activity, while microbial PMEs are less sensitive to their ionic environment (Rexova-Benkova and Markovic 1976).

Plant PME activity increases with salt concentration up to an optimal concentration above which activity usually decreases. With NaCl, this optimal concentration is ca. 0.2 M (Hultin et al. 1966; Lee and MacMillan 1968; Brady 1976; Lourenco and Catutani 1984; Lin et al. 1989). The nature of the anion might also intervene in this stimulation. Salt addition allows a shift of the pH optimum towards acidic values (Lin et al. 1989; Moustacas et al. 1991). It has been proposed that positive charges would interact with free carboxyl groups which would otherwise trap the enzyme (since plant PMEs have basic pIs, and are therefore cationic even at slightly basic pHs). Indeed, pectin de-esterification leads to the appearance of negative charges which can interact with PMEs, especially at acidic pHs. Monovalent cations would then compete with cationic proteins for fixation on pectate, thus allowing the liberation of PMEs. However, this is probably not the sole mechanism of activation, since some "acidic" vegetal PMEs are also activated by cations (Lin et al. 1989; Komae et al. 1990).

The effect of divalent cations on PME activity appears to be more complex than in the case of monovalent cations. At low concentrations (5–25 mM), calcium strongly enhances PME activity; whereas at higher concentrations, possibly due to the formation of pectate gel, it decreases the reaction rate. In *Vigna radiata*, magnesium is a more potent activator than calcium. Manganese can equally activate both free and fixed PME, while magnesium and calcium activate the fixed enzyme, and only weakly the solubilized one (Goldberg et al. 1992b).

Monovalent or divalent cations, depending on their concentration, might also modify the affinity of PMEs for their substrate.

5.3 Structures

Several plant and microorganism PMEs have already been sequenced, either by means of protein sequencing (Markovic and Jörnvall 1986), or after sequencing of

PME genes. Several plant, bacterial and fungal PME genes have been cloned and sequenced, e.g. in tomato (Ray et al. 1988; Harriman et al. 1991), in *Brassica napus* (Albani et al. 1991), in *Arabidopsis thaliana* (Richard et al. 1994), in *Erwinia chrysanthemi* (Laurent et al. 1993), and in *Aspergillus niger* (Khanh et al. 1991). Plant and microorganism PME sequences can differ signicantly overall, but they share several identical or similar regions, which likely participate in the structure of the enzyme-active site (Laurent et al. 1993). Sequence identity between plant PMEs is usually high in their C-terminal end, whereas their N-terminal sequences vary a lot between species (Richard et al. 1994).

Because only few studies have been reported on PME sequences, there is little information on their structural features. Inhibition of tomato PME activity by iodine led to the conclusion that a tyrosine residue(s) was present at the catalytic site. Later, Markovic and Jörnvall (1986), after the entire proteinaceous sequencing of one PME isoform from tomato, proposed that some tyrosine and histidine residues were probably implied at the catalytic site of the enzyme. Lastly, the occurrence of two intrachain disulphide bridges was demonstrated in tomato pectin esterase (Markovic and Jörnvall 1992).

6 Roles of PMEs

6.1 Plant Pathogens

Plant pathogenic microorganisms (fungi and bacteria) secrete several cell wall digesting enzymes, including cellulases, pectin and pectate lyases (PL), endo- and exo-polygalacturonases and PMEs. It has been shown that PME activity facilitates pectin degradation by polygalacturonases (PG), the latter being less active on methylated pectins than on acidic ones. These fungal and bacterial PMEs differ both structurally (M_r, pI, sequence) and enzymically (specific activity, K_m, V_{max}, pH sensitivity, thermosensitivity, etc.) from plant PMEs (Rexova-Benkova 1976). They often exhibit acidic pIs, higher specific activity and are much less sensitive to pH. Secretion of the cell wall degrading enzymes by these plant pathogens is frequently dependent upon the environmental composition. Pectinolytic enzymes are usually inducible (Bateman and Millar 1966); for example, in calli of *Orobanche aegyptiaca*, a dicotyledonous plant root parasite, the presence of pectin in the culture medium induces an earlier secretion of pectinolytic enzyme (BenHod et al. 1993).

PME alone cannot be the sole factor responsible for tissue maceration and hypersensitive responses, but probably acts in a synergistic way with other cell wall degrading enzymes. Recently, studies have shown that filtrates from phytopathogenic cultures contain secreted substances capable of solubilizing plant cell wall fragments that kill plant cells. These substances might be responsible for the hypersensitive response of plants. Several pectinolytic enzymes, including PMEs, were present in the plant cell-killing fractions purified from these filtrates;

nevertheless, mixtures of purified pectic enzymes (PME, PG and PL) alone could not kill plant cells (Bucheli et al. 1990). In all cases, it is not conceivable that PME alone is responsible for such complex processes as tissue maceration or a hypersensitive response.

6.2 Fruit Maturation

Koch and Nevins (1989) studied the degree of methylation of pectins during tomato fruit maturation. They observed an important decrease in DE during the ripening of tomato fruit, from 90% DE at the mature green stage to 35% DE at the ripe stage. They showed that both PME and PG (polygalacturonase) were expressed during maturation, PME being expressed before PG. As in phytopathogenic processes, they proposed that the action of PME enhanced PG activity (PG being less active on methylated pectins than on acidic pectins). PG action would in turn lead to softening of the cell wall by subsequent hydrolysis of the pectins. Also, PME activity has been monitored in other plant species during fruit maturation, and has been reported to increase during fruit maturation of tomato, highbush blueberries, and papaya (Catutani and Lourenco 1982; Pressey and Avants 1982; Proctor and Miesle 1991). In *Ficus awkeotsang* Makino achenes, Lin et al. (1990) showed that PME can be induced in the pericarp of the achenes by exposure of the achenes to air. Harrimann et al. (1991) studied the expression of a tomato pectin methyl esterase gene in several ripening mutants of tomato. Their work led to the conclusion that PME expression in the tomato fruit pericarp is highly regulated through mRNA synthesis and stability, protein stability, and delayed protein synthesis. Recently, Tieman et al. (1992) have obtained transgenic tomato plants transformed with an antisense PME construct. These plants were able to fructify and undergo normal fruit maturation, but the pectic composition of their fruits appeared to be greatly modified compared to control plants. As could be expected, their pectins were much more methylated, and the average size of pectic molecules was significantly greater, probably due to a decreased efficiency of PG action. Nevertheless, one must consider that these results were obtained on a material where several PME isoforms had been reported, and it cannot be excluded that the transgenic plants could express other isoforms to balance the effect of the antisense mRNA, which corresponded to only one isoform. Gaffe et al. (1994) have recently demonstrated that only one group of PME isoforms is actually inhibited by the antisense gene present in these transgenic tomatoes.

6.3 Cell Elongation

Due to their role in fruit maturation and their modifying activity in the cell wall structure, it has been proposed that PMEs might play a role in plant cell elongation (Moustacas et al. 1986; Nari et al. 1986). On the one hand, PMEs produce acidic pectins, which could then interact with cations (especially calcium) present in the cell wall, consequently reinforcing the wall structure. On the other hand, PME action produces protons, i.e. a decrease in cell wall pH. It has been proposed that this acidifying action could regulate the activity of other cell wall enzymes, like glucanases or polygalacturonases, likely to be implicated in the hydrolysis of cell wall components. Indeed, it has long been known that these cell wall degrading enzymes need acidic pHs to work. Recently, slightly more complex models for plant cell growth regulation via PME activity have been proposed (Moustacas et al. 1991; Charnay et al. 1992).

These somewhat opposing consequences of PME action could be explained by considering that, in plants, several PME isoforms are often present in the cell wall, depending on the development stage of the organs. Some of these isoforms might demethylate pectins blockwise, resulting in the stiffening of the wall if calcium links these acidic pectic chains. In contrast, other isoforms might demethylate pectins in a random way, thus providing action sites for cell wall depolymerizing enzymes. Only a study of the action pattern of the different PME isoforms present in the cell wall will confirm or contradict this hypothesis.

7 Conclusions and Perspectives

From all the data that can be found in the literature, it appears that it is very difficult to determine precisely the role of PMEs in plants, since their activity parameters differ greatly when they are fixed or solubilized. A better understanding of this problem might be achieved by investigating PME, thus addressing, synthesis processes and the way in which PMEs are exported to the cell wall, together with studies on the positioning of cell wall polysaccharides inside the apoplasm. Also, molecular biology studies might be helpful, by giving more information on PME primary and secondary structures, and by allowing one to follow their expression, for example with northern blot analysis.

It is evident that PMEs take part in a multi-factorial regulation system, and the only way to approach this system is to elaborate models of increasing complexity, by modifying the different parameters one after another. To conclude, this work, which began a long time ago, still bears the promise of new and interesting results that will lead us to a better understanding of the complex phenomena that take place in the plant cell wall.

References

Albani D, Altosaar I, Arnison PG, Fabijanski SF (1991) A gene showing sequence similarity with pectin esterase is specifically expressed in developing pollen of *Brassica napus*. Sequences in its 5′ flanking region are conserved in other pollen-specific promoters. Plant Mol Biol 16:501–513

Bateman D, Millar RL (1966) Pectic enzymes in tissue degradation. Annu Rev Phytopathol 4:119–146
BenHod G, Losner D, Joel DM, Mayer AM (1993) Pectin methylesterase in calli and germinating seeds of *Orobanche aegyptiaca*. Phytochemistry 32:1399–1402
Bertheau Y, Madgidi-Hervan E, Kotoujanski A, Nguyen The C, Coleno A (1984) Detection of depolymerase isoenzymes after electrophoresis. Anal Biochem 139:383–389
Bordenave M, Goldberg R (1993) Purification and characterization of pectin methylesterases from mung bean hypocotyl cell walls. Phytochemistry 33:999–1003
Bordenave M, Goldberg R (1994) Immobilized and free apoplastic pectinmethylesterases in mung bean hypocotyl. Plant Physiol 106:1151–1156
Borrego F, Tari M, Manjon A, Iborra JL (1989) Properties of pectinesterase immobilized on glycophase-coated controlled-pore glass. Appl Biochem Biotechnol 22:129–140
Brady CJ (1976) The pectinesterase from the pulp of the banana fruit. Aust J Plant Physiol 3:163–172
Bucheli P, Doares SH, Albersheim P, Darvill A (1990) Host-pathogen interactions. XXXVI. Partial purification and characterization of heat-labile molecules secreted by the rice blast pathogen that solubilize plant cell wall fragments that kill plant cells. Physiol Mol Plant Pathol 36:159–173
Catutani AT, Lourenco EJ (1982) Pectinesterase activity in ripening papaya. An Farm Quim São Paulo 22:3–10
Charnay D, Nari J, Noat G (1992) Regulation of plant cell-wall pectin methyl esterase by polyamines - interactions with the effects of metal ions. Eur J Biochem 205:711–714
Evans R, MacHale D (1978) Multiple forms of pectinesterase in limes and oranges. Phytochemistry 17:1073–1075
Forster H (1988) Pectinesterases from *Phytophtora infestans*. Methods Enzymol 161:355–361
Gaffe J, Tieman DM, Handa AK (1994) Pectin methylesterase isoforms in tomato (*Lycopersicon esculentum*) tissues. Effects of expression of a pectin methylesterase antisense gene. Plant Physiol 105:199–203
Gaffé J, Morvan C, Jauneau A, Demarty M (1992) Partial purification of flax cell wall pectin methylesterase. Phytochemistry 31:761–765
Giovane A, Quagliuolo L, Castaldo D, Servillo L, Balestrieri C (1990) Pectin-methylesterase from *Actinidia chinensis* fruit. Phytochemistry 29:2821–2823
Glasziou KT (1959) The localization and properties of pectin-methylesterase of *Avena* coleoptiles. Physiol Plant 12:670–680
Goldberg R (1984) Changes in the properties of cell wall pectin methylesterase along the *Vigna radiata* hypocotyl. Physiol Plant 61:58–63
Goldberg R, Morvan C, Roland JC (1986) Composition, properties and localization of pectins in young and mature cells of the mung bean hypocotyl. Plant Cell Physiol 27:417–429
Goldberg R, Bordenave M, Pierron M, Prat R, Mutaftschiev S (1992a) Enzymatic processes in growing cell walls, possible control by pectinmethylesterases. In: Masuda Y, ed Plant cell walls as biopolymers with physiological functions. Yamada Science Foundation, Osaka, pp 269–274
Goldberg R, Pierron M, Durand L, Mutaftschiev S (1992b) In vitro and in situ properties of cell wall pectinmethylesterases from mung bean hypocotyl. J Exp Bot 43:41–46
Hagerman AE, Austin PJ (1986) Continuous spectrophotometric assay for plant pectin methyl esterase. J Agric Food Chem 34:440–444
Harriman RW, Tieman DM, Handa AK (1991) Molecular cloning of tomato pectin methylesterase gene and its expression in Rutgers, Ripening Inhibitor, Non Ripening and Never Ripe tomato fruits. Plant Physiol 97:80–87
Hultin HO, Sun B, Bulger J (1966) Pectin-methylesterase of the banana. Purification and properties. J Food Sci 31:320–327
Jarvis MC (1984) Structure and properties of pectin gels in plant cell walls. Plant Cell Environ 7:153–164
Khanh NQ, Ruttkowski E, Leidinger K, Albrecht H, Gottschalk M (1991) Characterization and expression of a genomic pectin methyl esterase-encoding gene in *Aspergillus niger*. Gene 106:71–77

Koch JL, Nevins DJ (1989) Tomato fruit cell wall. I. Use of purified tomato polygalacturonase and pectinmethylesterase to identify developmental changes in pectins. Plant Physiol 91:816–822

Komae K, Sone Y, Kakuta M, Misaki A (1990) Purification and characterization of pectinesterase from *Ficus awkeotsang*. Agric Biol Chem 54:1469–1476

Kotoujanski A (1987) Molecular genetics of pathogenesis by soft-rot Erwinias. Annu Rev Phytopathol 25:405–430

Laurent F, Kotoujanski A, Labesse G, Bertheau Y (1993) Characterization and overexpression of the *pem* gene encoding pectin methylesterase of *Erwinia chrysanthemi* strain 3937. Gene 131:17–25

Lee M, MacMillan JD (1968) Mode of action of pectic enzymes. I. Purification and properties of tomato pectinesterase. Biochemistry 7:4005–40010

Lim YM, Chung MCM (1993) Isolation and characterization of pectin methylesterase from papaya. Arch Biochem Biophys 307:15–20

Lin TP, Liu CC, Chen SW, Wang WY (1989) Purification and characterization of pectinmethylesterase from *Ficus awkeotsang* Makino achenes. Plant Physiol 91:1445–1453

Lin TP, Liu CC, Huang RS, Wang WY, Feng TY (1990) Induction of pectinmethylesterase in the pericarp of achenes of the jelly-fig *Ficus awkeotsang* Makino. Plant Cell Physiol 31: 533–537

Lourenco EJ, Catutani AT (1984) Purification and properties of pectinesterase from papaya. J Sci Food Agric 35:1120–1127

Ma R, Reese JC, Black IV WC, Bramel-Cox P (1990) Detection of pectinesterase and polygalacturonase from salivary secretions of living greenbugs, *Schizaphis graminum* (Homoptera: Aphididae). J Insect Physiol 36:507–512

MacMillan GP, Johnston DJ, Morel JB, Pérombelon MCM (1993) A pH-independant assay for pectin methyl esterase for use in column chromatography. Anal Biochem 209:377–379

Markovic O, Jörnvall H (1986) Pectinesterase. The primary structure of the tomato enzyme. Eur J Biochem 158:455–462

Markovic O, Jörnvall H (1992) Disulfide bridges in tomato pectinesterase: variations from pectinesterase of other species; conservation of possible active site segments. Protein Sci 1:1288–1292

Markovic O, Kohn R (1984) Mode of pectin deesterification by *Trichoderma reesei* pectinesterase. Experientia (Basel) 40:842–843

Markovic O, Machova E, Slezarik A (1983) The action of tomato and *Aspergillus foetidus* pectinesterases on oligomeric substrates esterified with diazomethane. Carbohydr Res 116:105–111

Markovic O, Machova E, Dankova L, Slezarik A (1988) Pectinesterase and polygalacturonase from alfalfa (*Medicago sativa* L.). Biologia (Bratislava) 43:1069–1076

Moustacas AM, Nari J, Diamantidis G, Noat G, Crasnier M, Borel M, Ricard J (1986) Electrostatic effects and the dynamics of enzyme reactions at the surface of plant cells. 2. The role of pectin methylesterase in the modulation of electrostatic effects in soybean cell walls. Eur J Biochem 155:191–197

Moustacas AM, Nari J, Borel M, Noat G, Ricard J (1991) Pectin methylesterase, metal ions and plant cell-wall extension. The role of metal ions in plant cell-wall extension. Biochem J 279:351–354

Nari J, Noat G, Diamantidis G, Woudstra M, Ricard J (1986) Electrostatic effects and the dynamics of enzyme reactions at the surface of plant cells. 3. Interplay between limited cell-wall autolysis, pectin methylesterase activity and electrostatic effects in soybean cell walls. Eur J Biochem 155:199–202

Nari J, Noat G, Ricard J (1991) Pectin methylesterase, metal ions and plant cell-wall extension. Hydrolysis of pectin by plant cell-wall pectin methylesterase. Biochem J 279:343–350

Pressey R, Avants JK (1992) Multiple forms of pectin esterase in tomatoes. Phytochemistry 11:3139–3142

Pressey R, Woods FM (1992) Purification and properties of two pectinesterases from tomatoes. Phytochemistry 31:1139–1142

Proctor A, Miesle TJ (1991) Polygalacturonase and pectinmethylesterase activities in developing highbush blueberries. Hortscience 26:579–581

Ray J, Knapp J, Grierson D, Bird C, Schuch W (1988) Identification and sequence determination of a cDNA clone for tomato pectinesterase. Eur J Biochem 174:119–124

Rexova-Benkova L, Markovic O (1976) Pectic enzymes. Adv Carbohydr Chem Biochem 33:323–385

Richard L, Qin LX, Gadal P, Goldberg R (1994) Molecular cloning and characterization of a putative pectin methylesterase cDNA in *Arabidopsis thaliana* (L.). FEBS Lett 355:135–139

Rillo L, Castaldo D, Giovane A, Servillo L, Balestrieri C, Quagliuolo L (1992) Purification and properties of pectin-methylesterase from mandarin orange fruit. J Agric Food Chem 40:591–593

Roeb L, Stegemann H (1975) Pectin-methylesterases: isozymes, properties and detection in polyacrylamide gels. Biochem Physiol Pflanz 168:607–615

Sakellaris G, Evangelopoulos AE (1989) Production, purification, and characterization of extracellular pectinesterase from *Lactobacillus plantarum* (strain BA11). Biotechnol Appl Biochem 11:503–507

Schejter A, Marcus L (1988) Isoenzymes of pectinesterase and polygalacturonase from *Botrytis cinerea*. Methods Enzymol 161:366–373

Schink B, Zeikus JG (1983) Characterization of pectinolytic enzymes of *Clostridium thermosulfurogenes*. FEMS Microbiol Lett 17:295–298

Seymour GB, Lasslett Y, Tucker GA (1987) Differential effects of pectolytic enzymes on tomato polyuronides in vivo and in vitro. Phytochemistry 26:3137–3139

Seymour TA, Preston JF, Wickler L, Lindsay JA, Wei CI, Marshall MR (1991) Stability of pectinesterase of Marsh white grapefruit pulp. J Agric Food Chem 39:1075–1079

Sheiman MI, MacMillan JD, Miller L, Chase T Jr (1976) Coordinated action of pectinesterase and polygalacturonate lyase complex of *Clostridium multifermentans*. Eur J Biochem 64:565–572

Solms J, Deuel H (1955) Über den Mechanismus der enzymatischen Verseifung von Pektinstoffen. Helv Chim Acta 38:321–329

Spok A, Stubenrauch G, Schorgendorfer K, Schwab H (1991) Molecular cloning and sequencing of a pectinesterase from *Pseudomonas solanacearum*. J Gen Microbiol 137:131–140

Tieman DM, Harriman RW, Ramamohan G, Handa AK (1992) An antisense pectin methylesterase gene alters pectin chemistry and soluble solids in tomato fruit. Plant Cell 4:667–679

Tucker GA, Robertson NG, Grierson D (1982) Purification and changes in activities of tomato pectin-esterase isoenzymes. J Sci Food Agric 33:396–400

Van Rijssel M, Gerwig GJ, Hansen TA (1993) Isolation and characterization of an extracellular glycosylated protein complex from *Clostridium thermosaccharolyticum* with pectin methylesterase and polygalacturonate hydrolase activity. Appl Environ Microbiol 59:828–836

Versteeg C (1979) Pectinesterases from the orange fruit – their purification, general characteristics and juice-cloud destabilizing properties. PhD Thesis, Agric Univ Wageningen, Netherlands

Versteeg C, Rombouts FM, Pilnik W (1978) Purification and some characteristics of two pectinesterase isoenzymes from orange. Lebensm Wiss Technol 11:267–274

Woods PJ, Siddiqui IR (1971) Determination of methanol and its application to measurements of pectin ester content and pectin methylesterase activity. Anal Biochem 39:418–428

Zhang GF, Staehelin LA (1992) Functionnal compartmentation of the Golgi apparatus of plant cells. Plant Physiol 99:1070–1083

Zimmerman RE (1978) A rapid assay for pectinesterase activity which can be used as a prescreen for pectinesterase inhibitors. Anal Biochem 85:219–223

Probing the Subunit Composition and Topology of Plasma Membrane-Bound (1,3)-β-Glucan (Callose) Synthases

AAZBRUCE P. WASSERMAN, XIAOQUN QI, LUCILLE M. BARONE and AYONG WU

1 Introduction

The plasma membrane (PM) is one of the most important, but least understood, membrane systems in higher plants. The PM serves as the gateway to the cell, and possesses numerous transport, recognition, biosynthetic, and containment functions. In higher plants, the PM serves as the site of synthesis of cellulose [(1,4)-β-linked glucan] microfibrils, which are a virtually ubiquitous component of plant cell walls. It is the site of synthesis for wound-induced callose [(1,3)-β-linked glucan], which is an essential part of plant defense responses. Developmentally regulated callose synthesis, as occurs in cotton fibers, sieve plates, plasmodesmata, and pollen tubes, is also thought to occur at the PM.

While cellulose is synthesized by the intact plant in abundant quantities, incubation of isolated membranes with uridine diphosphate glucose (UDPGlc) generally yields high levels of callose. Although synthesis of low levels of cellulose has now been reported in cotton (Okuda et al. 1993; Kudlicka et al. 1995), callose remains the predominant in vitro product in most plant systems examined. Two hypotheses have been proposed to explain this phenomenon. One is that the PM contains distinct callose and cellulose synthase complexes. In this scenario, the cellulose synthase is extremely labile. Accordingly, upon cellular disruption, the capacity of membranes to synthesize cellulose is lost, therefore callose becomes the predominant enzyme product. A second hypothesis proposes that cellulose and callose synthase activities are present within a single enzyme complex (Jacob and Northcote 1985; Delmer 1987). Upon disruption, this complex is deregulated in such a way that callose, rather than cellulose, is produced. In both models, substrate is thought to be delivered to the enzyme complex at the cytoplasmic face of the PM, and layers of fibrillar product are deposited at the extracellular space as glucan synthase complexes move through the plane of the lipid bilayer. Thus, each respective complex is postulated to contain a pore or channel through which nascent glucan is translocated across the membrane. Morphological studies documenting the existence of particle complexes within the PM provide support for this model (Mueller and Brown 1980; Reiss et al. 1984; Herth 1985; Giddings and Staehelin 1988).

Red beet (*Beta vulgaris* L.) storage tissue has proven to be an excellent source of callose synthase, due to the enzyme's abundance and relative stability. For more than a decade, we have conducted a comprehensive biochemical dissection of the

Modern Methods of Plant Analysis, Vol. 17
Plant Cell Wall Analysis
Edited by H.-F. Linskens and J.F. Jackson

beet callose synthase complex (Sloan et al. 1987; Wu et al. 1991; Wasserman et al. 1992; Wu and Wasserman 1993; Qi et al. 1995). Thus, it has been possible to systematically transform this system from what was once a "black box" to a defined set of polypeptides to which we now have antibodies and cDNA clones. To briefly summarize, these efforts yielded: (1) elucidation of the topology of the callose synthase complex within the PM and derivation of a working model (Wasserman et al. 1992; Wu and Wasserman 1993: Fig. 1). (2) Production of a highly purified callose synthase complex (Wu and Wasserman 1993: Fig. 2). (3) Identification of a set of polypeptides (31, 29 and 27kDa) that strongly copurify with callose synthase activity (Fig. 2) as PM-localized members of the major intrinsic protein (MIP) or aquaporin family (Qi et al. 1995). The current working model (Fig. 1) proposes that callose synthase is a multisubunit complex containing a catalytic subunit (possibly the 57-kDa species) closely associated with transmembrane PM intrinsic proteins (PMIPs), forming a channel through which nascent glucans are translocated. The close association of PMIPs with callose synthase activity has led us to speculate that PMIPs serve as the channels for translocation of nascent β-glucan polymer across the plane of the PM (Qi et al. 1995).

Detailed protocols describing the assay and solubilization of higher plant callose synthases were published in Volume 10 of this series (Wasserman et al. 1989). The primary objective of this chapter is to provide detailed descriptions of recently adopted protocols used to identify callose synthase polypeptides in red beet storage tissue, and to explore their topology within the plant PM. Moreover, improvements to previously described protocols are described.

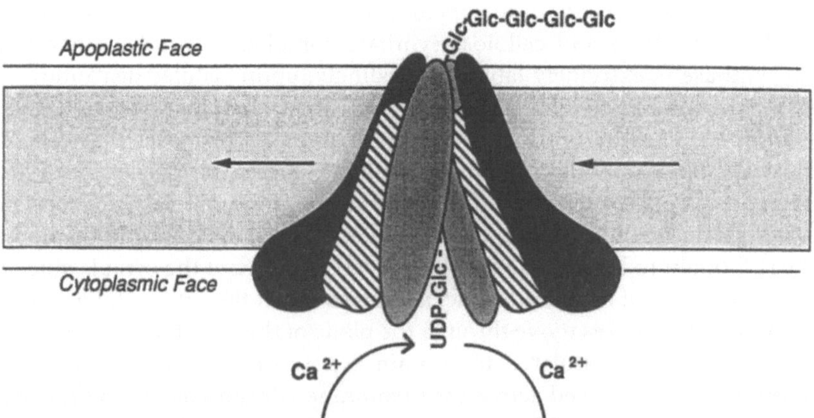

Fig. 1. Biochemically-derived working model of the callose synthase complex from *Beta*. The *arrows* indicate the direction of movement of the enzyme complex as it diffuses through the PM. For illustrative purposes, catalysis of one molecule of UDPGlc with one polymer strand of callose is shown. Morphologically derived models (Mueller and Brown 1980; Reiss et al. 1984; Herth 1985; Giddings and Staehelin 1988) indicate that each cellulose microfibril contains multiple glucan chains derived simultaneously from a single complex

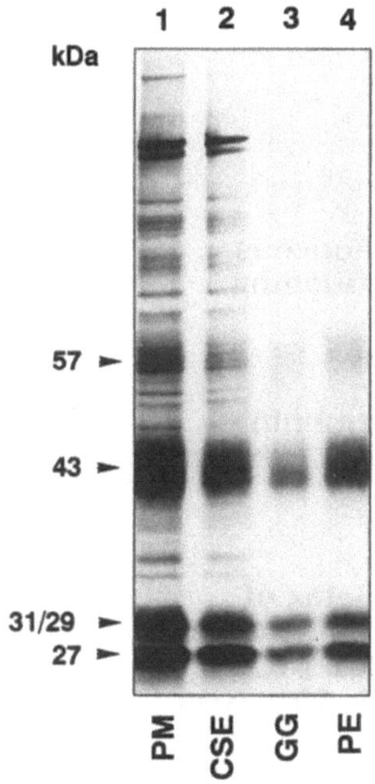

Fig. 2. SDS-PAGE of PMs and fractions enriched in PMIPs and callose synthase activity. The *PM* fraction was prepared by aqueous two-phase partitioning (*lane 1*) and proteins were solubilized using CHAPS (*CSE*, *lane 2*). Further purification consisted of glycerol gradient centrifugation (*GG*, *lane 3*) followed by product entrapment (*PE*, *lane 4*). The SDS sample buffer contained 50 mM β-mercaptoethanol, but no DTT. Polypeptide molecular masses are indicated. Polypeptides were visualized by silver staining following enhancement by Coomassie Blue (Daiichi double-staining protocol, Integrated Separation Systems, Enprotech, Hyde Park, MA)

2 Membrane Isolation, Enzyme Assay, and Solubilization

Membrane isolation procedures are summarized in Fig. 3. Aqueous two-phase partitioning (Widell et al. 1982; Morre and Morre 1989) has gradually become the method of choice for preparing PM fractions, replacing discontinuous sucrose gradient centrifugation. Although yields of purified PM prepared by two-phase partitioning are tenfold lower (2.2 vs. 24 μg PM protein per g FW), callose synthase recovery is only reduced by two- to threefold (28 vs. 74%) (Wu et al. 1991). As a consequence, callose synthase-specific activity is enriched 18-fold by aqueous partitioning as opposed to 3.5-fold by sucrose gradient centrifugation. Furthermore, PM produced by aqueous two-phase partitioning is free of tonoplast contamination, as judged by the absence of the 55-kDa ATP-binding subunit of the tonoplast ATPase. Vesicle fractions enriched in apoplastic side-out and cytoplasmic side-out PM are prepared from enriched PM using a freeze/thaw method (Palmgren et al. 1990).

2.1 Isolation of Crude Membrane Fractions

Red beets are obtained locally. The tops are removed and the roots may be stored in moist vermiculite for up to 3 weeks at 4 °C. Microsomal membranes are isolated by differential centrifugation (Wasserman et al. 1985, 1989; Wasserman and

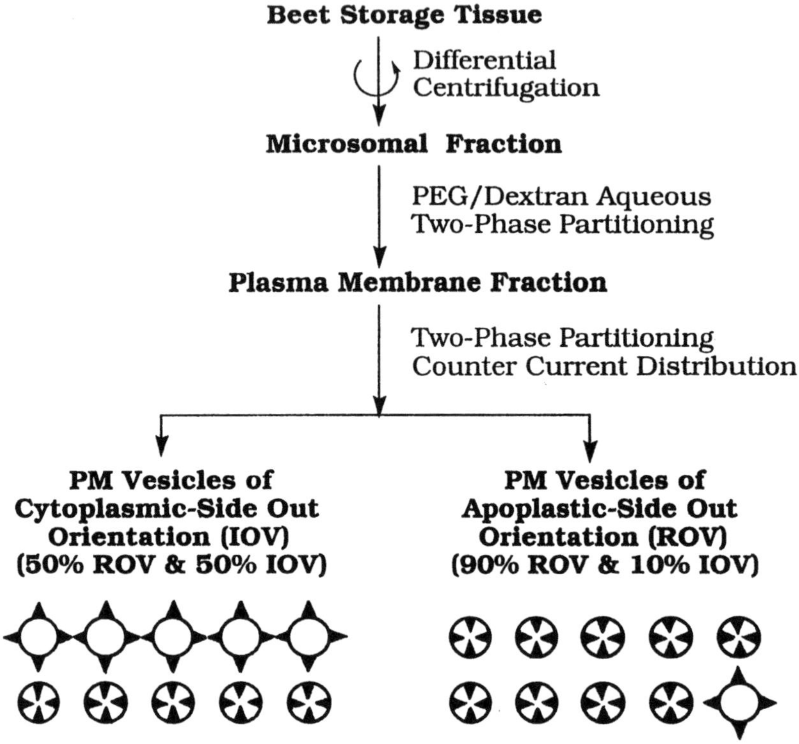

Fig. 3. Membrane isolation scheme. IOV and ROV denote inside out and right-side out vesicles, respectively

McCarthy 1986) as follows: plant tissue is surface sterilized in 0.5% sodium hypochlorite for 10 min. After rinsing with distilled water, the tissue is peeled, diced into 2–5 cm blocks, and homogenized in 250 mM sucrose, 3 mM ethylenediaminetetraacetic acid (EDTA), 0.5% polyvinylpyrolidone-40, 1 mM dithiothreitol (DTT), and 70 mM N-2-hydroxyethylpiperazine-N′-2-ethanesulfonic acid (HEPES)/NaOH, pH 8 at a ratio of 50 ml buffer per 100 g fresh weight tissue by a single pass through a juice extractor. If large capacity GSA and Ti 45 (Beckman, Palo Alto, CA) rotors are available, 1500 g of tissue may be processed per day. All steps are conducted at 4 °C.

Extracts are filtered through two layers of Miracloth to remove residual particulate matter. Mitochondria and other large particles are then removed by centrifugation for 15 min at 13 000 × *g*. Microsomal membranes are recovered by centrifuging the supernatant for 45 min at 80 000 × *g*. Microsomes are then resuspended to a protein concentration of 3–6 mg ml^{-1} in storage buffer consisting of 1 mM DTT, 1 mM HEPES/NaOH, pH 7.2, 15% (v/v) glycerol and may be stored at −80 °C for at least several months. Protein content is determined by the Coomassie Blue dye-binding method (Bradford 1976) using bovine serum albumin as the protein standard.

2.2 Plasma Membrane Isolation

To prepare PM by aqueous two-phase partitioning, two sets of six 50-ml centrifuge tubes are prepared. The following components are then added: 10.2 g of 20% dextran T 500, 5.1 g of 40% polyethylene glycol (PEG) (MW 3350), and 0.8 ml of 0.4 M tris[hydroxymethyl]aminomethane hydrochloride (Tris)-HCl, pH 7.5. Four ml of microsomal membranes is added to one set of tubes; 4 ml of water is added to the parallel set. The tubes are mixed by inverting 30–40 times. The tubes are then centrifuged at 1300 × *g* for 5 min in a TJ-6 centrifuge (Beckman). Following centrifugation, the upper PEG layers are withdrawn from the parallel set and discarded. The PEG layers from the membrane-containing set are added to the dextran layers from the parallel set, and the tubes are again centrifuged. The PEG layers (120 ml) are then diluted with 2 vol of resuspension buffer (250 mM sucrose, 1 mM DTT, 10 mM HEPES/NaOH, pH 7.5) and centrifuged at 80 000 × *g* for 80 min using a Ti 45 rotor. The pellet fractions containing PM are resuspended in storage buffer (Sect. 2.1), assayed for protein, and stored at –80 °C.

2.3 Membranes of Defined Sidedness: Inside Out and Right-Side Out Vesicles

The starting vesicle fraction (predominantly apoplastic side out) is subjected to four freeze/thaw cycles (with cycle temperatures of –80 and 20 °C, respectively) to produce vesicles of mixed orientation (Palmgren et al. 1990; Wu and Wasserman 1993). Vesicles of cytoplasmic side-out orientation are then separated from the apoplastic side-out vesicle fraction by phase partitioning using a three-step countercurrent distribution. This is done by adding 6 ml of freeze/thawed PM vesicles (4 mg protein) to a 26-g phase mixture (32 g total), with polymers at a final concentration of 6%. After a three-step phase partitioning procedure, an enriched cytoplasmic side-out PM vesicle fraction is recovered from the lower phase of tube 1. An apoplastic side-out PM vesicle fraction is obtained from the upper phase of tube 3. The dextran phase of tube 1 and the PEG phase of tube 3 are diluted ten- and threefold, respectively, in resuspension buffer (Sect. 2.2). The cytoplasmic side-out and apoplastic side-out-enriched vesicle fractions are collected by centrifugation (80 000 × *g* for 80 min), resuspended in the above buffer containing 1 mM EDTA, assayed for activity and protein, and stored at –80 °C.

2.4 Callose Synthase Assay

Enzyme assays are generally conducted in 100-µl assay mixtures in 1.5-ml capped polypropylene centrifuge tubes containing substrate, appropriate cofactors, buffer, and membranes. Concentrations of individual components will vary somewhat depending upon the tissue being analyzed (Wasserman et al. 1989). Red beet callose synthase assay mixtures of 100 µl typically contain 2 mM $CaCl_2$, 5 mM $MgCl_2$, 5 mM cellobiose, 0.01% digitonin, 1 mM UDP[^{14}C]Glc (0.1 mCi per mmol), and 50 mM Tris-HCl, pH 7.4. Protein levels may vary between 0.5 and 50 µg, depending upon the relative purity of the sample to be assayed. To mixtures of 90 µl containing all components except substrate, reactions are initiated by addition of 10 µl of UDP[^{14}C]Glc. The mixtures are incubated at 30 °C for 5 min, however, times may be extended for samples of dilute activity such as gradient fractions. Reactions are terminated by heating at 90 to 100 °C in a water-filled Dri-Bath for 5 min.

To isolate radiolabeled ethanol-insoluble glucan, reaction mixtures are cooled and spotted in their entirety onto Whatman GF/A filters. Sample numbers are penciled onto the filters before spotting. The filters are dried under a heat lamp and positioned onto a 12.5-cm Buchner funnel. The filters are then washed successively with 66% (v/v) ethanol containing 0.85 mM EDTA, 66% ethanol, and 70% ethanol. The filters are washed with acetone, dried, placed in scintillation vials with 5 ml of Liquiscint (National Diagnostics, Somerville, NJ) and counted. Quenching is corrected by counting known amounts of UDP[^{14}C]Glc that have been spotted on filters and dried before counting. An additional set of filters is spotted with known amounts of UDP[^{14}C]Glc. These are extracted with ethanol as above to determine wash efficiency and background binding of UDP[^{14}C]Glc for reaction blanks. One enzyme unit is defined as the nmol of glucose incorporated into ethanol-insoluble glucan per min.

2.5 Callose Synthase Solubilization

Using a two-step solubilization procedure, the specific activity of solubilized enzyme may be approximately doubled by a preliminary membrane extraction to remove contaminating proteins. The first step, which suppresses callose synthase solubilization, utilizes a low concentration of 3-[(3-cholamidopropyl) dimethylammonio]-1-propane sulfonate (CHAPS) (0.3% w/v) detergent and incorporates 5 mM Mg^{2+}. The low concentration of CHAPS removes peripheral proteins, while Mg^{2+} aggregates the enzyme causing it to sediment at 100 000 × *g*. Following removal of the 100 000 × *g* supernatant, callose synthase is then solubilized using 0.6% CHAPS and 1 mM each of ethylene glycol-bis-(β-aminoethylether)N,N,N′,N′-tetraacetic acid (EGTA) and EDTA (Sloan et al. 1987).

To solutions containing microsomal membranes at protein concentrations of 2 to 3 mg ml^{-1} or plasma membranes at 1 to 2 mg ml^{-1}, an equal volume of buffer containing 0.6% (w/v) CHAPS, 10 mM $MgCl_2$ and 100 mM Tris/HCl, pH 7.5, is added. The suspensions are incubated on ice for 15 min and centrifuged at 80 000 × *g*. Pellets are resuspended to a concentration of 2 to 3 mg ml^{-1} protein in 50 mM HEPES/NaOH, pH 7.5, and 15% glycerol and to this is added an equal volume of solubilization buffer consisting of 1.2% CHAPS, 2 mM EDTA, 2 mM EGTA, and 100 mM Tris/HCl, pH 7.5. The suspensions are incubated on ice for 30 min and centrifuged as before.

A one-step solubilization using CHAPS may suffice for certain applications. To solutions containing membranes at protein concentrations of 2 to 3 mg ml^{-1}, an equal volume of buffer containing 1.2% (w/v) CHAPS, 2 mM EDTA, 2 mM EGTA, and 100 mM Tris/HCl, pH 7.5, is added dropwise on ice with stirring. The suspensions are incubated for 30 min on ice and then centrifuged at 80 000 × *g* in a Type 40 rotor (Beckman) for 30 min. The supernatants are recovered for further use. Pellets are resuspended in 0.6% (w/v) CHAPS, 1 mM EDTA, 1 mM EGTA, and 50 mM Tris/HCl, pH 7.5, and assayed. CHAPS-solubilized enzyme requires the presence of digitonin (0.01%) in assay mixtures for activity to be expressed (Sloan et al. 1987).

Percent solubilization is defined as the ratio of activity recovered in the supernatant to the sum of activities recovered in the supernatant and resuspended pellet. Stability of activity to the solubilization process may be quantitated by taking the ratio of total activity recovered (sum of activities in the supernatant and resuspended pellet) to the total activity added to the solubilization mixture.

3 Callose Synthase Topology, Purification, and Subunit Composition

3.1 Polypeptide Depletion

When purified enzyme is unavailable, polypeptide depletion represents a way to correlate an enzyme activity with specific membrane-bound proteins and to distinguish whether a membrane protein is integral or peripherally bound. The principle of polypeptide depletion is to selectively extract polypeptides (using detergents or chaotropic agents) from a purified or partially purified enzyme complex, and to determine if loss of activity correlates with depletion of specific polypeptides. This technique has gained wide use for characterizing the polypeptide composition of multisubunit, membrane-bound enzymes such as tonoplast (Rea et al. 1987; Lai et al. 1988; Bowman et al. 1989; Moriyama and Nelson 1989) and coated vesicle (Arai et al. 1988; Adachi et al. 1990b) ATPases, and the chloroplast ATP synthase (Feng and McCarty 1990a,b; Avital and Gromet-Elhanan 1991).

Prior to the combined use of glycerol gradient centrifugation and product entrapment, polypeptide depletion was used to help narrow down the list of candidate components of the beet callose synthase complex (Wasserman et al. 1992). This study produced two major findings. First, many polypeptides were ruled out as potential callose synthase subunits. We were able to narrow the number of candidate polypeptides (i.e., 83, 57, 43, 31, 29, and 27 kDa). Moreover, this study demonstrated the strongly hydrophobic nature of each of these integral membrane proteins.

To conduct polypeptide depletion, PM (90 µg of protein in 300 µl of 50 mM Tris/HCl, pH 7.5) is incubated with either a chaotrope (500 mM KSCN, 500 mM KI, and 100 mM Na_2PO_4) or a detergent (0.15% Triton X-100 or 1% octyl glucoside) for 20 min at 4 °C. The suspensions are then centrifuged at 40 000 × *g* for 30 min, and the pellets resuspended in 150 µl Tris/HCl, pH 7.5. Pellets and supernatants are assayed for residual activity and analyzed for polypeptide composition by SDS-PAGE (Wasserman et al. 1992).

3.2 Vectorial Proteolysis

Limited proteolysis utilizing PM vesicles of defined sidedness has been extensively used to determine the orientation of membrane-bound proteins (Davis and Hammes 1989; Adachi et al. 1990a; Imajoh-Ohmi et al. 1992; Marty et al. 1992). Using trypsin and other proteases to probe cytoplasmic side-out and apoplastic side-out membranes, we demonstrated that protease-sensitive sites on the callose synthase complex are located primarily at the cytoplasmic face of the PM (Wu and Wasserman 1993). During the course of this work, we also found that callose synthase activity is largely (75%) preserved upon treatment of PM vesicles with Pronase E. While numerous polypeptides including at least four previously implicated as possible callose synthase components (92, 83, 57, and a broad ca. 43 kDa band) were extensively hydrolyzed, polypeptides of 31, 29 and 27 kDa resisted Pronase E hydrolysis (Fig. 4). Following solubilization and purification, however,

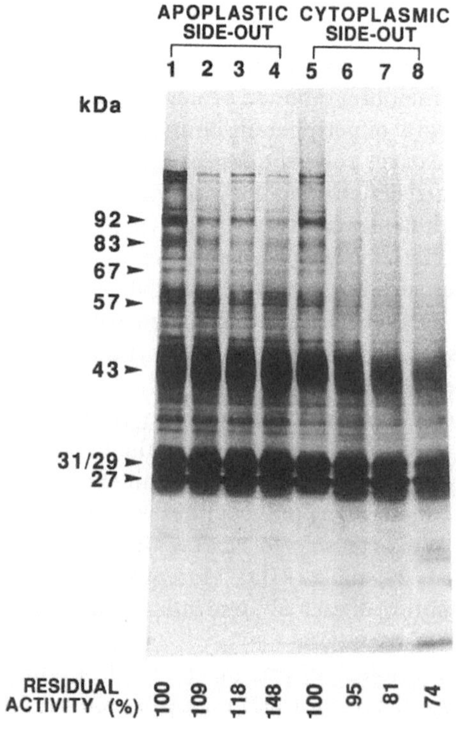

Fig. 4. Effect of Pronase E on callose synthase activity and polypeptide profiles. Reaction mixtures enriched in vesicles of either apoplastic (*lanes 1–4*) or cytoplasmic side-out (*lanes 5–8*) orientation (45 µg of protein in 120 µl reaction mixtures) were treated with increasing levels of Pronase E (0–40 µg ml^{-1}) for 20 min at room temperature. Pronase E levels are as follows; *lanes 1* and *5*, controls; *lanes 2* and *6*, 10 µg ml^{-1}; *lanes 3* and *7*, 20 µg ml^{-1}; *lanes 4* and *8*, 40 µg ml^{-1}. Samples were assayed and prepared for SDS-PAGE as described here and in Wu and Wasserman (1993)

callose synthase was rapidly hydrolyzed by Pronase E and trypsin. This indicated that most protease-sensitive sites on callose synthase are deeply embedded within the PM.

Controlled hydrolysis of sided vesicles by trypsin is conducted as follows: time-dependent inactivation of callose synthase is determined using incubation mixtures of 0.72 ml containing 180 µg of apoplastic side-out or cytoplasmic side-out PM vesicles, 18 µg of trypsin, and 50 mM Tris/HCl, pH 7.5. The mixtures are held at room temperature and at each selected time point, four 20-µl aliquots are withdrawn and added to 70 µl of assay mixture components (as above) containing 5.4 µg of soybean trypsin inhibitor. Since trypsin is rapidly autolyzed under these incubation conditions, fresh trypsin is added (to a level of 25 $\mu g\,ml^{-1}$) at 30-min intervals. To determine latency duplicate aliquots are assayed in the absence and presence of 0.01% digitonin. Callose synthase assays are then initiated by the addition of 10 µl of 10 mM UDP[^{14}C]Glc and are conducted as described above.

Concentration-dependent loss of callose synthase activity is determined in incubation mixtures of 60 µl containing 18 µg of membrane vesicles, with trypsin levels ranging from 0 to 112 $\mu g\,ml^{-1}$, in 50 mM Tris/HCl, pH 7.5. After incubation for 20 min, 10-µl aliquots are withdrawn and assayed as above.

To analyze the polypeptide composition of trypsin-treated vesicles, incubation mixtures of 240 µl are centrifuged at 40 000 × *g* for 40 min. The pellets are resuspended in 100 µl of 250 mM sucrose, 1 mM DTT, 1 mM EDTA, and 10 mM HEPES, pH 7.5. Sample buffer (28 µl) is then combined with an aliquot of resuspended pellet (12 µl), and electrophoresis conducted. When using other proteases (e.g., Pronase E) (Wu and Wasserman 1993), protease levels and incubation times must be adjusted accordingly.

3.3 Callose Synthase Purification

Chromatographically, CHAPS- or digitonin-solubilized callose synthase behaves more like a supramolecular complex than a typical detergent-solubilized enzyme. Due to its size and hydrophobicity, callose synthase activity is difficult to recover from gel filtration or ion-exchange columns. On the other hand, centrifugation-based techniques such as glycerol gradient centrifugation and product entrapment provide effective means for obtaining highly enriched enzyme.

3.3.1 Glycerol Gradient Centrifugation

Solubilized callose synthase (0.75 ml) is layered on 4.2 ml 25 to 40% (w/v) linear glycerol gradients containing 0.5% CHAPS, 3 mM EDTA, 3 mM EGTA, and 50 mM Tris/HCl, pH 7.5. The gradients are centrifuged at 200 000 × *g* for 4 h, and are then fractionated into 14 tubes of 0.34 ml each. Pellets are resuspended in 0.34 ml of 2 mM DTT, 1 mM EDTA, 250 mM sucrose, and 10 mM HEPES, pH 7.2 in 15% glycerol. Fractions are assayed for enzyme activity, protein levels, and polypeptide composition. Fractions at the leading edge of the activity peak and the pellet usually contain the highest specific activities of callose synthase, and are characteristically enriched in polypeptides of 92, 57, 43, 31, 29, and 27 kDa (Fig. 5). Replacement of chelators in glycerol solutions by divalent cations (1 mM $CaCl_2$,

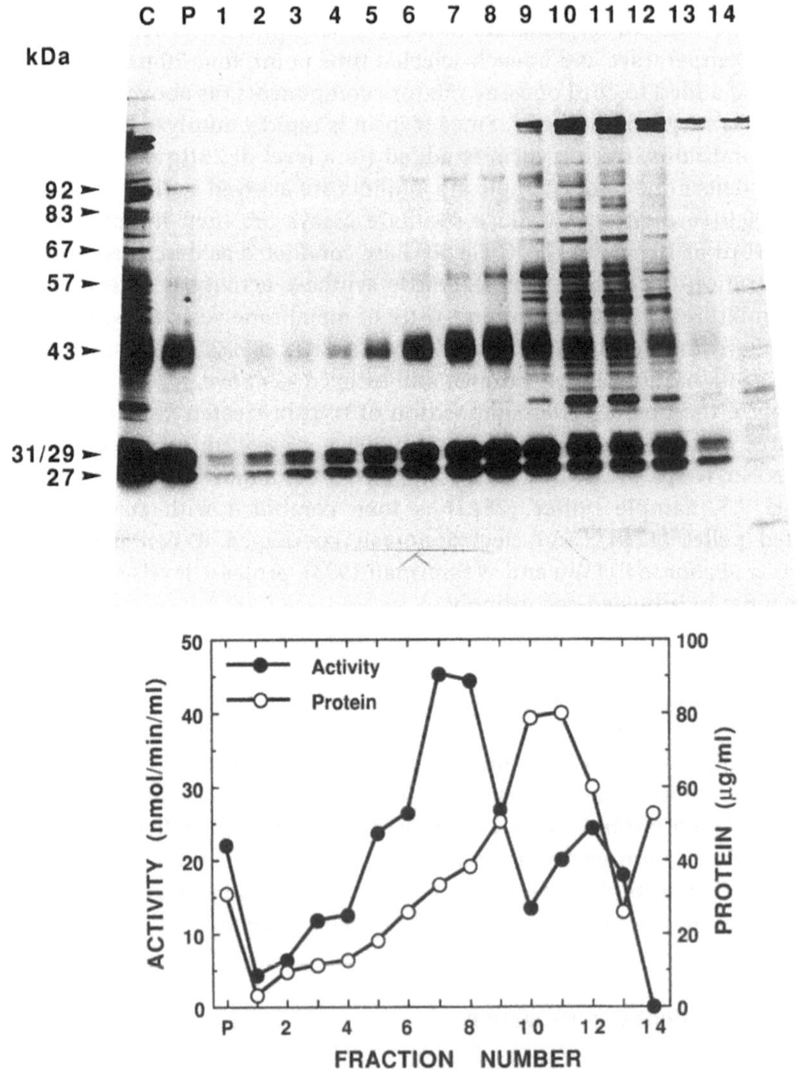

Fig. 5. Glycerol gradient centrifugation of polypeptides solubilized from plasma membranes by CHAPS. A CHAPS extract (165 µg protein in 0.75 ml) was layered onto a 25–40% (w/v) glycerol gradient and then centrifuged for 4 h at 200 000 × *g* in an SW 50.1 rotor. Fractions (340 µl) were collected and analyzed for activity (•), protein (○), and polypeptide composition by SDS-PAGE (484) *C* and *P* denote CHAPS-Solubilized callose synthase and gradient pellet, respectively

5 mM $MgCl_2$) will cause enzyme activity to shift towards gradient pellets (Wu et al. 1991).

3.3.2 Product Entrapment

In the standard procedure adopted for obtaining time- and substrate-dependent product entrapment (Wu et al. 1991), reaction mixtures ranging from 1 to 30 ml are prepared by combining 1 vol of solubilized enzyme (0.2–0.4 mg protein ml^{-1}) with 4 vol of an effector mix to give final concentrations of: 1 mM $CaCl_2$, 0.5 mM $MgCl_2$, 0.2 mM EDTA, 0.2 mM EGTA, 0.02% digitonin, 5 mM cellobiose, 1.5% glycerol, and 0.12% (w/v) CHAPS in 50 mM Tris/HCl, pH 7.5 (EDTA, EGTA, and CHAPS levels reflect carryover from the soubilization buffer). Synthesis of glucan is initiated by addition of 1 mM UDPGlc. After 40 min at 30 °C, reactions are stopped by cooling to 4 °C for 30 min, and mixtures are centrifuged at low speed (7000 × *g* for 40 min at 4 °C in a Sorvall SA-600 rotor or 11 000 × *g* for 40 min at 8 °C in a microcentrifuge). Supernatants (S1) are carefully removed. Pellets (P1) are resuspended in 0.5 to 15 ml of 0.3% (w/v) CHAPS, 3 mM EDTA, 10% glycerol, and 50 mM Tris/HCl, pH 7.5, and recentrifuged at 7000 or 11 000 × *g* for 40 min. The resulting supernatants (S2) are carefully withdrawn and pellets (P2) are resuspended as above, but with 0.05% CHAPS. Each fraction is assayed for enzyme activity and protein. Pellets are most efficiently resuspended by sonicating for several seconds in 40 to 50 μl of resuspension buffer at room temperature before bringing to volume.

4 Biochemical Characterization of Integral Plant Plasma Membrane Proteins

4.1 Characterization of Disulfide-Linked Aggregate Formation

Highly enriched callose synthase from red beet consists of strongly hydrophobic polypeptides of 43, 31, and 27 kDa (Fig. 2). The 31- and 27-kDa proteins have been shown to be plant PM members of the MIP superfamily of proteins (Qi et al. 1995). Integral membrane proteins have a tendency to form aggregates under oxidizing conditions (Johnson et al. 1990; Hofte et al. 1992; Maeshima 1992). Thus, it was not surprising that proportions of the three species observed on SDS gels vary depending on the levels of reducing agent present in SDS-PAGE sample buffers. In the absence of DTT, the 43-kDa species predominates, whereas DTT added to 100 mM causes almost complete conversion of the 43-kDa species to the PMIPs 31 and 27 (Fig. 6). The broad band at 43 kDa is therefore a disulfide-linked aggregate composed of PMIPs 31 and 27.

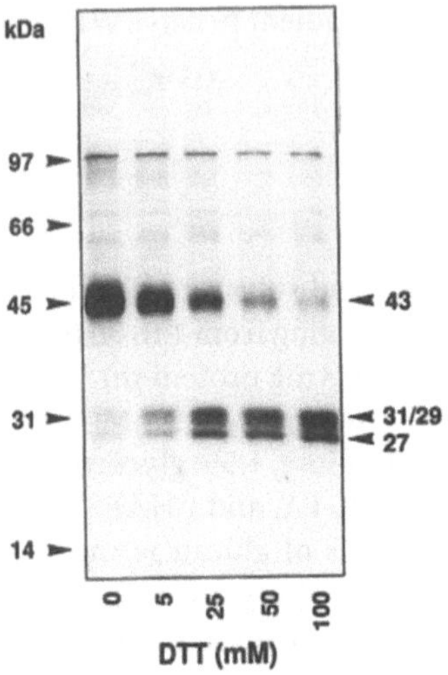

Fig. 6. Concentration-dependent conversion of the 43-kDa species by DTT. The PMIP-enriched fraction was isolated from a CHAPS extract by glycerol-gradient centrifugation. Aliquots (5 μg) were combined with various levels of DTT and were incubated at 30 °C for 15 min. SDS sample buffer prepared without β-mercaptoethanol was added and heated at 45 °C before electrophoresis. Following incubation at 45 °C for 5 min, the samples were electrophoresed (Qi et al. 1995)

4.2 Recovery of Hydrophobic Plant-Derived Membrane Proteins for Sequencing and Antibody Production

With difficult to purify plant membrane enzyme complexes, direct protein sequencing of individual polypeptides provides essential information for protein identification, and for the design of molecular probes such as oligonucleotide primers for PCR.

Recovery of the proteins consists of SDS-PAGE, excision of the bands of interest, and electroelution. Membrane caps are presoaked in protein elution buffer (192 mM glycine, 0.1% SDS, 25 mM Tris, pH 7.5) at 60 °C for 1 h. The gel slices are loaded into glass tubes and are eluted at 60 mA (8–10 mA/tube) for 4 h (Model 422 Electro-Eluter, Bio-Rad, Richmond, CA). To limit aggregation, reducing agent must be present throughout. Electroeluted samples are concentrated (Centricon 10 microconcentrators, Amicon, Beverly, MA) and protein levels are quantified by SDS-PAGE.

For sequencing, samples are electrophoresed and blotted onto polyvinylidene difluoride membranes (Bio-Rad) using 10 mM 3-[cyclohexylamine]-1-propanesulfonic acid (CAPS) in 30% methanol, pH 11. Blots are stained with 0.5% Ponceau S in 1% acetic acid for 10 s and destained with 1% acetic acid. Bands of interest are then excised for either N-terminal sequencing or for preparation of internal fragments. Amino acid composition can be determined directly from the electroeluted proteins.

Sequence information for PMIPs 31 and 27 was obtained by in situ digestion by trypsin, with products separated by HPLC. Protein sequences are then provided by standard procedures, with sensitivity to approximately 100 pmol. Using this procedure, the longest sequence obtained from PMIPs 31 and 27, were 19, and 23 amino acids, respectively (Qi et al. 1995). Amino acid sequence information is used to scan databases such as MacDNASIS Pro (Hitachi Software Engineering

America, Ltd.) or the blast network service (National Center for Biotechnology Information).

Similarly, for antibody production, SDS gels are run using PM fractions as the antigen source. After staining with Coomassie Blue, bands corresponding to the antigen are excised and electroeluted as described above. Elution occurs at constant current (60 mA) for 4 h. The electroeluted protein is used to immunize New Zealand white rabbits. Membrane proteins whose aggregated state may result in less surface exposure and thus poor antigenicity can be fragmented into smaller peptides prior to injection (Englehard 1994). Proteins are incubated with trypsin attached to polyacrylamide beads (1 unit trypsin μg^{-1} protein) (Sigma, St. Louis, MO) at 30 °C for 30 min with shaking. Solutions are then centrifuged at $13\,000 \times g$ to remove the trypsin and supernatants are collected for injection. SDS gels are run to confirm digestion of the antigen into smaller molecular mass peptides. In addition, Coomassie Blue-stained gel slices are minced and used directly for immunization. Approximately 100 µg total of both gel slices and trypsinized protein was used for generating antiserum against PMIP27.

4.3 Antibody Characterization

As expected, polyclonal antibodies raised against PMIP27 recognize the 43- and 27-kDa species present in callose synthase-enriched PM fractions (Fig. 7). Not surprisingly, crude antiserum also contains antibodies recognizing other proteins of unknown identity (e.g., a 35-kDa polypeptide). To achieve the desired specificity, anti-PMIP27 is further purified by affinity purification.

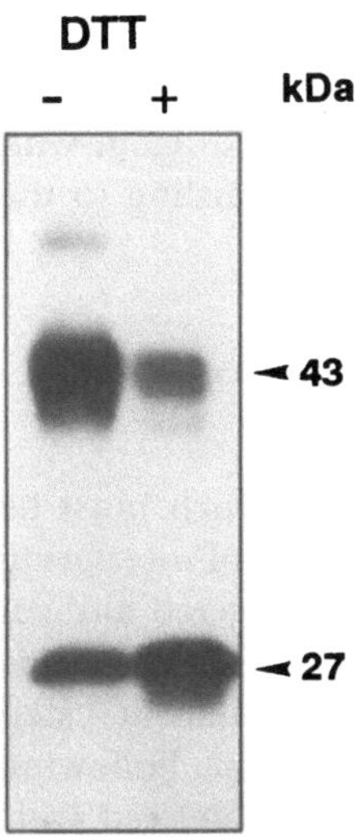

Fig. 7. Immunoblot probed with affinity-purified anti-PMIP27. PM prepared by aqueous two-phase partitioning was electrophoresed in the absence or presence of 100 mM DTT, transferred to a nitrocellulose membrane, and probed with affinity-purified anti-PMIP27. Bands were visualized by ECL

4.3.1 Immunoblotting

SDS-PAGE is first carried out followed by electrotransfer to nitrocellulose membranes (Schleicher and Schuell, Keene, NH). The gel and membrane are each equilibrated for 10 min in transfer buffer (0.02% SDS, 25 mM Tris, 160 mM glycine, pH 8.7). The transfer sandwich (consisting of support pads, filter paper, transfer membrane, and the gel) is assembled and inserted into a transfer tank (Hoefer Scientific, San Francisco, CA), the membrane facing the positive electrode. Transfer is completed after 2-4 h at 63 V and 7 °C.

Visualization is then conducted using either alkaline phosphatase-labeled secondary antibodies (Harlow and Lane 1988), or by the enhanced chemiluminescence (ECL) technique when increased sensitivity is required. For visualization using alkaline phosphatase-conjugated secondary antibodies, membranes are removed from transfer sandwiches and incubated with Tris-buffered saline (TBS) (14 mM NaCl, 25 mM Tris/HCl, pH 7.5), containing 1% BSA for 2–3 h at room temperature with shaking. Antibody is then added to appropriate dilutions, and held for 2 h. Membranes are then washed several times with TBS, before adding 10 µl of goat-anti-rabbit IgG conjugated to alkaline phosphatase (BioRad) in 30 ml of TBS (1:3000 dilution). After 2 h, secondary antibody is removed and membranes are washed with TBS containing 1% Triton X-100 and 5 mM EDTA, pH 7.5 for 10 min. Membranes are equilibrated in developing buffer (100 mM NaCl, 5 mM $MgCl_2$, 100 mM Tris/HCl, pH 9.5) and nitro-blue tetrazolium and 3-bromo-4-chloroindolyl phosphate are added. Color development is terminated by rinsing in distilled water.

Visualization by ECL (Amersham, Arlington Heights, IL) is conducted as follows: antigen-containing membranes are soaked in TBST buffer (150 mM NaCl, 0.1% Tween 20, 10 mM Tris/HCl, pH 7.4) containing 1% BSA. After 1 h, blots are washed several times with TBST and are then incubated for 1 h with primary antibody in TBST. Due to the improved sensitivity of this method, it is often necessary to increase antibody dilution relative to colorimetric visualization methods. Membranes are washed with TBST, incubated with goat-anti-rabbit IgG conjugated to horseradish peroxidase (Amersham) for 1 h, and washed again with TBST. Blots are then developed using ECL detection reagents according to the manufacturer's protocol.

4.3.2 Affinity Purification of Anti-PMIP27

Crude immune serum often contains contaminating antibodies which must be removed. Blot-affinity purification is a simple method to rid serum of interfering antibodies which may cross-react with proteins other than the targeted antigen (Tang 1993). Affinity purification utilizes antigen-containing nitrocellulose strips to capture and purify antibodies recognizing a specific polypeptide. In the first step SDS-PAGE is run, loading approximately 20 µg of antigen per gel. Following electrotransfer to nitrocellulose, horizontal strips corresponding to desired antigens are excised from the membrane. One lane is usually set aside for staining with 0.5% Ponceau S in 1% acetic acid in order to locate specific antigens. Strips are incubated in 10 ml TBST, pH 8, containing 1% BSA for 30 min, and are then rinsed with TBST. Antiserum is added (1 : 100) to TBST containing 0.05% BSA and 0.025% NaN_3, the strips are incubated overnight, and then washed with TBST.

To elute captured antibodies, nitrocellulose strips are cut into small pieces and placed in tubes containing 2 ml of elution buffer (500 mM NaCl, 0.05% Tween-20, 100 mM glycine, pH 2.8) for 3 min over ice with gentle mixing. Eluted antibodies are then transferred to tubes containing 0.3 ml of neutralization buffer (1 M Tris/HCl, pH 8.1). The elution-neutralization procedure is repeated, and the strips then rinsed with 2 ml TBST for 1 min. The two eluates and the TBST rinse are then combined, and 0.75 ml of 10% BSA and 75 µl of 5% NaN_3 are added. Diluted affinity-purified antibodies are analyzed for their specificity by Western blotting (Sect. 4.3.1), and can be stored at 4 °C. A desirable feature of the affinity-purification procedure is that both crude serum and the nitrocellulose strips are reusable. The nitrocellulose strips are neutralized with TBST, dried, and stored at 4 °C for future use. Incubation serum can be stored at 4 °C in NaN_3 and can be reutilized until desired antibodies are depleted.

Figure 7 illustrates PM (in the absence and presence of DTT) probed using affinity-purified anti-PMIP27 with visualization by ECL. Exclusive recognition of PMIP27 and the 43-kDa species demonstrates that affinity-purification provides antibodies of the desired high specificity.

5 Summary

Procedures used to characterize membrane-bound enzymes are well established, however, the polysaccharide synthases found in the PM of higher plants represent a unique challenge. In addition to catalysis of glycosyl transfer, these synthases must be associated with subunits that facilitate translocation of nascent polymer through the plane of the PM. The co-purification of putative channel-forming proteins such as PMIPs 31 and 27 with callose synthase is consistent with the existence of membrane pores capable of facilitating polymer translocation, however, much work remains to be done in order to estabish whether a definitive link exists between PMIPs and the callose biosynthetic process. The availability of highly specific antibodies and cDNA clones provides the means to pursue these objectives. Prior to this work, the only integral PM protein from plant sources for which sequence information had been obtained via direct purification of a polypeptide from isolated membranes was the 100-kDa H^+-ATPase, which is responsible for generation of proton gradients across the membrane (Sussman 1994).

Identification and cloning of callose synthase subunits should also help to resolve long-standing questions regarding relationships between callose and cellulose synthesis. For example, the availability of amino acid sequences and cDNA for callose synthase subunits will reveal whether sequence homologies exist with known bacterial cellulose synthase genes, and therefore define the extent to which callose and cellulose synthases are structurally related. This information will also help to establish the extent to which callose and cellulose-synthesizing complexes share a common subunit structure, and facilitate the use of genetic approaches to examine cell wall biogenesis. Continued development and refinement of techniques used to isolate and analyze integral membrane proteins should greatly advance our knowledge of the mechanism by which cell wall polymers are assembled.

Acknowledgments. Support was provided in part by the National Science Foundation (Grants MCB 92-05832 and the New Jersey Agricultural Experiment Station with State and Hatch Act Funds. and MCB 95-07266)

References

Adachi I, Arai H, Pimental R, Forgac M (1990a) Proteolysis and orientation on reconstitution of the coated vesicle proton pump. J Biol Chem 265:960–966

Adachi I, Puopolo K, Marquez-Sterlong N, Arai H, Forgac M (1990b) Dissociation, cross-linking, and glycosylation of the coated vesicle proton pump. J Biol Chem 265:967–973

Arai H, Terres G, Pink S, Forgac M (1988) Topography and subunit stoichiometry of the coated vesicle proton ATPase. J Biol Chem 263:8796–8802

Avital S, Gromet-Elhanan Z (1991) Extraction and purification of the β-subunit and an active ab-core complex from the spinach chloroplast CF_0F_1-ATP synthase. J Biol Chem 266:7067–7072

Bowman BJ, Dschida WJ, Harris T, Bowman EJ (1989) The vacuolar ATPase of *Neurospora crassa* contains an F_1-like structure. J Biol Chem 264:15606–15612

Bradford MM (1976) A rapid and sensitive method for the quantitation of microgram quantities of protein using the principles of protein dye-binding. Anal Biochem 72:248–254

Davis CB, Hammes GG (1989) Topology of the yeast plasma membrane proton-translocating ATPase. J Biol Chem 264:370–374

Delmer DP (1987) Cellulose biosynthesis. Annu Rev Plant Physiol 38:259–290

Englehard VH (1994) How cells process antigens. Sci Am 271, 2:54–61

Feng Y, McCarty RE (1990a) Chromatographic purification of the chloroplast ATP synthase (CF_0-CF_1) and the role of the CF_0 subunit IV in proton conduction. J Biol Chem 265:12474–12480

Feng Y, McCarty RE (1990b) Purification and reconstitution of active chloroplast F_0. J Biol Chem 265:5104–5109

Giddings TH, Staehelin LA (1988) Spatial relationship between microtubules and plasma-membrane rosettes during the deposition of primary wall microfibrils in *Closterium* sp. Planta 173:22–30

Harlow E, Lane D (1988) Antibodies, a laboratory manual. Cold Spring Harbor, New York

Herth W (1985) Plasma-membrane rosettes involved in localized wall thickening during xylem vessel formation of *Lepidium sativum* L. Planta 164:12–21

Hofte H, Hubbard L, Reizer J, Ludevid D, Herman EM, Chrispeels MJ (1992) Vegetative and seed-specific forms of tonoplast intrinsic protein in the vacuolar membrane of *Arabidopsis thaliana.* Plant Physiol 99:561–570

Imajoh-Ohmi S, Tokita K, Ochiai H, Nakamura M, Kanegasaki S (1992) Topology of cytochrome b_{558} in neutrophil membrane analyzed by anti-peptide antibodies and proteolysis. J Biol Chem 267:180–184

Jacob SR, Northcote DH (1985) In vitro glucan synthesis by membranes of celery petioles: the role of the membrane in determining the type of linkage formed. J Cell Sci (Suppl) 2:1–11

Johnson KD, Herman EM, Chrispeels MJ (1990) An abundant, highly conserved tonoplast protein in seeds. Plant Physiol 91:1006–1013
Kudlicka K, Brown RM, Li L, Lee JH, Shin H, Kuga S (1995) β-Glucan synthesis in the cotton fiber. IV. In vitro assembly of the cellulose I allomorph. Plant Physiol 107:111–123
Lai S, Randall SK, Sze H (1988) Peripheral and integral subunits of the tonoplast H^+-ATPase from oat roots. J Biol Chem 263:16731–16737
Maeshima M (1992) Characterization of the major integral protein of vacuolar membrane. Plant Physiol 98:1248–1254
Marty I, Brandolin G, Gagnon J, Brasseur R, Vignais PV (1992) Topography of the membrane-bound ADP/ATP carrier assessed by enzymatic proteolysis. Biochemistry 31:4058–4065
Moriyama Y, Nelson N (1989) H^+-translocating ATPase in Golgi apparatus. J Biol Chem 264:18445–18450
Morre DJ, Morre DM (1989) Preparation of mammalian plasma membranes by aqueous two-phase partition. BioTechniques 7:946–958
Mueller SC, Brown RM, Jr (1980) Evidence for an intramembrane component associated with a cellulose microfibril-synthesizing complex in higher plants. J Cell Biol 84:315–326
Okuda K, Kudlicka K, Koga S, Brown RM, Jr (1993) β-Glucan synthesis in the cotton fiber. I. Identification of β-1,4- and β-1,3-glucans synthesized in vitro. Plant Physiol 101:1131–1142
Palmgren MG, Askerlund P, Fredrikson K, Widell S, Sommarin M, Larsson C (1990) Sealed inside-out and right-side-out plasma membrane vesicles. Optimal conditions for formation and separation. Plant Physiol 92:871–880
Qi X, Tai CY, Wasserman BP (1995) Plasma membrane intrinsic proteins (PMIPs) of *Beta vulgaris* L. Plant Physiol 108:387–392
Rea PA, Griffith CJ, Manolson MF, Sanders D (1987) Irreversible inhibition of H^+-ATPase of higher plant tonoplast by chaotropic anions: evidence for peripheral location of nucleotide-binding subunits. Biochim Biophys Acta 904:1–12
Reiss H-D, Schnepf E, Herth W (1984) The plasma membrane of the *Funaria* caulonema tip cell: morphology and distribution of particle rosettes, and the kinetics of cellulose synthesis. Planta 160:428–435
Sloan ME, Rodis P, Wasserman BP (1987) CHAPS solubilization and functional reconstitution of β-glucan synthase from red beet root. Plant Physiol 85:516–522
Sussman MR (1994) Molecular analysis of proteins in the plant plasma membrane. Annu Rev Plant Physiol Mol Biol 45:211–234
Tang WJY (1993) Blot-affinity purification of antibodies. In: Asai DJ (ed) Antibodies in cell biology Academic Press, San Diego, pp 95–104
Wasserman BP, McCarthy KJ (1986) Regulation of plasma membrane β-glucan synthase by phospholipids. Reactivation of Triton X-100 extracted glucan synthase by phospholipids. Plant Physiol 82:396–400
Wasserman BP, Ventola CL, Eiberger LL (1985) Glucan biosynthesis in red beet root microsomes and tissue slices. J Agril Food Chem 33:144–149
Wasserman BP, Wu A, Harriman, RW (1992) Probing the molecular architecture of (1,3)-β-glucan (callose) synthase: polypeptide depletion studies. Biochem Soc Trans 20:18–22
Wasserman BP, Frost DJ, Lawson SG, Mason TL, Rodis PS, Sabin RD, Sloan ME (1989) Biosynthesis of cell wall polysaccharides: membrane isolation, in vitro glycosyl transferase assay and enzyme solubilization. In: Linskens HF, Jackson JF (eds) Modern methods of plant analysis, vol 10. Plant fibers. Springer Berlin Heidelberg New York, pp 1–11
Widell S, Lundborg T, Larsson C (1982) Plasma membranes from oats prepared by partition in an aqueous polymer two-phase system. On the use of light-induced cytochrome b reduction as a marker for the plasma membrane. Plant Physiol 70:1429–1435
Wu A, Wasserman BP (1993) Limited proteolysis of (1,3)-β-glucan (callose) synthase from *Beta vulgaris* L.: topology of protease-sensitive sites and polypeptide identification using Pronase E. Plant J 4:683–695
Wu A, Harriman RW, Frost DJ, Read SM, Wasserman BP (1991) Rapid enrichment of CHAPS-solubilized UDP-glucose: (1,3)-β-glucan (callose) synthase from *Beta vulgaris* L. by product entrapment. Entrapment mechanisms and polypeptide characterization. Plant Physiol 97:684–692

Subject Index

Absidia glauca 85
Acer 69, 77
acidic xylans, fractionation 34
Actinidia
- *chinensis* 170, 173
- *deliciosa* 46
active cell walls, isolation 49–51
adhesion/adhesive
- components, identification 105–110
- compounds 99
- definition 95
- molecules 97
- - purified 109
- mutants 105, 106
- perturbation 107, 108
- phase 99
- precursors 101
- process, polymerization polyphenols 102
- spatial development 106
- systems 104, 105
adhesion-reduced mutants 104
adhesiveness, gene control 97
affinity chromatography using lectins 9, 10
Agalinis purpurea 98
AGP (arabinogalactan proteins) 64
AIR (alcohol-insoluble residue) 19, 21
alcohol-insoluble residue (AIR) 19, 21
aleuron cell wall 77
algal-substratum adhesion 99, 100
alginate, adhesive mucilage 102
Allium sativum 31, 66
AMMS (anion micromembran suppressor) 152
- desalting 152, 153
ampholines, removement 15
anion-exchange chromatography
- separation of polysaccharides 8
- subfractionation of cell wall polymers 30
antibody-disruption experiments 109
Apedinella radians 100
apoplastic barrier 72
appressoria 98
aquaperin 182
Arabidopsis thaliana 126
arabinogalactan proteins (AGP) 64
Arachis hypogaea 14
Arum orientale 10
Aspergillus niger 171
assay for adhesions 105
autolysis
- incubations 53–55
- - procedure of 55
- products, cell wall analysis 56, 57
- reaction, optimization 54
- in vitro experiments 45

Beta bulgaris 181, 182
borate derivatives, affinity chromatography of mannans 10
Botrytis cinerea 171
Brassica napus 169
breathing, transient opening of pores 76
Brocchinia 66, 69, 73
buffers for autolysis experiments 47–69
bulk exclusion techniques, wall porosity 65, 67–71

calcium-mediated interactions 102
callose synthases 181–196
- assay 185, 186
- model 182
- purification 189–191
- solubilization 186–187
callosic walls 71–72
Candida albicans 90, 105
carbohydrates, lectin-plant, interactions 99
carboxyl group of pectic polysaccharides and CDTA with calcium, interactions, “loose complex” 27
carboxyl-reduction of methylated polysaccharides 37

Carica papaya 170, 173
Catenaria anguillulae 107
cation exchange chromatography, fractionation of plant polysaccharides 9
CDTA (cyclohexane-trans-1,2-diamine-NNN'N'-tetra acetate) 24, 27, 28
- determination by FTIR 39–41
- interaction with calcium 27
- pectin extraction 27, 28
cell
- elongation, role of PME 176–177
- structure, olive pulp 20
- wall (*see* wall)
cell-ECM adhesion 96
cell-substratum
- adhesion 96–100, 106
- glue 107
cellulosic residue (CR) 27, 29
Ceramium 99
Cercosporella 171
Chenopodium 69, 70
chitin
- biophysical identification 86–90
- from diatom spines 87, 88
- structure 81–82
- - assay 83
- - chitosome-entrapped 84
- - morphogenetic role 91
- - preparation 84, 85
- - synthase genes 82–92
- synthases, protease stimulation 82, 84
- synthesizing particles (*see* chitosomes)
- vectorial synthesis 84
- X-ray diffraction analysis 86, 87
chitinase digestion 85
chitosome-entrapped chitin synthases 84
chitosomes (chitin synthesizing particles) 84, 85, 90
Chlamydomonas 46
Chlorella 46
Chlostridium sterilium 46
Cicer arietinum 46
Citrus 170, 173
Ciucanu-Kerek methylation, polysaccharides 36–37
classes separation, cell wall constitutents 1
Clostridium 171
co-precipitation
- avoid 19
- phenolics 21
colleters 98
concanavalin A, prevents adehsion 103
conidiospores, adhesion 103–104
Coprinus cinereus 82
cotton fibre, MW distribution 12
CR (cellulosic residue) 27, 29
- supernatant of the dialysis of (sn-CR) 27
crude membrane fractions, isolation 183–184
CWM (cell wall material) 21
- composition, glycosylic linkage 38
- methylation analysis 37, 38
- preparation 21–24
- - from olive pulp 21–24, 25–29
- - of, solubilization of material 23, 24
cyclohexane-trans-1,2-diamine-NNNN-tetra acetate (CDTA) 24, 27, 28
cytorrhysis 67, 68
cytoskeletal-exocellular matrix, bridging 110

DCB (2,6-dichlorobenzonitrile), inhibitor of cellulose synthesis 74
depolymerization (DP) 151
desalting by AMMS 152–153
Dictyostelium 102, 110
digitonin, solublization of chitin synthases from membranes 84
disruption of hydrogen bonding between hemicellulose and cellulosic microfibrils 28
disulfide-linked aggregate formation, characterization 191
DP (depolymerization) 151
Drosera 98

ECL (enhanced chemiluminescence) 194
- visualization membranes by ECL 194, 195
ECM (extracellular matrix) 96–97, 99, 109
electroendosmotic effect, minimalization 11
electrophoresis, flat-bed isoelectric focusing 15
endo-PG (endo-polygalacturonase) 134
- depolymerization 137, 138, 139, 141
endocytosis, fluid phase 75
Enteromorpha 99
enzymes, wall-bound 1
EPS (extracellular polysaccharides), secretion 77

Erwinia chrysanthemi 169
ethanol
- graded precipitation with ethanol, cell wall polysaccharides 30
- heat inactivation, protection of pectin polymers 51, 52
- homogenization of cell walls 49, 50
- precipitation, cell wall autolysis products 56–57
exclusion of PEG 69–70
exocellular adhesive formation 101
expansins 12
extensin 12–15, 117–226
- expression of 125
- genes, mutations 126
- molecular characterization 122–123
- mutants 126
- promotor fusion with reporter genes 125
- in vivo localization 121
extracellular
- adhesive
- - material, extraction of 109
- sticky sites 97, 100

F-dextran 69, 72–74
- rate of movement 73
- uptake 69, 72, 74
FDA (fluorescein diacetate) 72
feruloyl esters of cell wall fragments, paper electrophoresis 11
Ficus 169, 170, 173
flat-bed isoelectric focusing electrophoresis 15
fractionation of acidic xylans 34
fragmentation cell walls, alteration micostructure 70
fruit maturation, role of PME 176
FTIR (Fourier Transform Infrared Spectroscopy) 39–41
- chitin 86–87
fucans, sulfated 102
Fucus 100–103, 105
- substrate adhesion 100–103
functional assays, transport of tracer molecules across cell wall 65–66
fungal phytopathogens, adhesion 98–99
Fusarium 103

gel filtration, pore size cell walls 70–71
gel permeation (*see* GPC)
genes encoding extensin 123–125
Gladiolus 66, 67, 68, 69, 72, 77
glass fiber paper electrophoresis, fractionation cell wall fragments 11
glucomanans 2
glucoronoarabinoxylan, esterification 77
glycerol gradient centrifugation, callose synthase 189–191
Glycine max 46, 69, 79, 170
glycoproteins
- in plant cell walls 12–15
- precipitation 13
glycosylic linkage composition of CWM 38
Gossypium hirsutum 56, 69
GPC (gel-permeation chromatography) 57, 130
- chromatography, cell wall autolysis products 57
- pectin 130
graded precipitation with ethanol, cell wall polysaccharides 30
GRP (glycine-rich proteins) 12

Hakomori methylation, polysaccharides 36
HCA (hydrophobic cluster analysis) 91
heat inactivation, in precence of ethanol, protection of pectin polymers 51–52
hemicellulose 2
- alkali-soluble effect of ionic form on separation 8
- degradation 28
- extracts 29
- polysaccharides, anion-exchange chromatography 32–34
high performance pressure system, separation of xyloglucan-derived oligosaccharides 7
hindered movement 75
histidin-rich extensin (HRGP) 13
homogeneity, moving boundary electrophoresis 10
homogenization
- of cell walls 49–50
- - in ethanol 52
- of tissues, to obtain cell walls 49–50
Hordeum 69, 74
Hormoseira 99
host surface, recognition 98
HPAEC (high-performance anion-exchange chromatography)
- analysis oligosaccharide 150–152
- sugar oligomers 151

HPAEC-MS
- characterization of, oligosaccharide 156–160
- maximum DP, oligosaccharide 153
- xylose oligomers 154
HPAEC-PAD (high-performance anion-exchange chromatography and pulsed amperometric detection) pectins 131–141
- mobile phase gradient 136
HPLC separations, pectin 130–134
HRGP (histidin-rich extensin) 13
hydrophobic plant-derived membrane proteins, recovery 192–193
hydroxyproline
- estimation 36
- rich glycoproteins 12

in vivo (*see* vivo)
in vitro (*see* vitro)
integral plant plasma membrane proteins, characterization 191–195
integrin-binding protein 102
ion exchange chromatography
- application to cell wall proteins 13–14
- fractionation cell wall polysaccharides 7–8
isoelectric focusing peroxydase isoenzymes 14–15
isolation of cell walls
- active cell walls 49–51
- inactive cell walls 46, 51–53
- by phenol-acetic acid-water 52

Lactobacillus plantarum 171
Laminaria 99
lectin-plant carbohydrates interactions 99
lignins 15–16
Lilium 71, 72
Linum usitatissimum 170, 173
"loose" complex, interactions of carboxyl group of pectic polysaccharides and CDTA with calcium 27
Lycopersicon esculentum 46, 70, 170, 173

macroconidial tip mucilage 103
Magnaporthe grisea 106
Malus 170
Medicago sativa 170
membrane
- isolation procedures 183–186
- proteins, antibody characterization 193–195
methylation analysis
- of CWM 37–38
- polysaccharides 36–39
microcapsules 70
microspheres, identification of sticky sites 100, 101
microspore development, callose 71–72
MIP (major intransic protein) 182
mobile phase gradient, HPAEC-PAD 136
Monilia laxa 171
Morinda 69, 73, 75
movement
- hindered, cell wall 75
- passing through cell wall 66
moving boundary electrophoresis, assessment of homogeneity 10
mucilage 99, 100
Mucor
- *hiemalis* 46
- *rouxii* 82
Musa acuminata 170
mutations, extensin genes 126
MW distribution of cotton fiber 12
Mytilus 97

NCP (non-cellulosic polysaccharides) 2, 16, 74
Nectria haematococca 103, 104, 106
Neurospora crassa 82, 91, 104
neutral sugars, analysis 35
Nicotiana 14, 69, 73, 74, 75, 77
non-cellulosic polysaccharides (*see* NCP) 2, 16, 74
non-melting flesh peaches 134–135
non-pectic polysaccharides 2

Olea europaea 19–41
oligogalacturonides 9
oligosaccharide
- HPAEC analysis 150–152
- - gradient program 149
- HPAEC-MS characterization 156–160
- ionization 153–155
- isolation from cell wall material 150
- maximum DP, HPAEC-MS 153
- sodiated 155
- xycloglucan-derived 6, 7
- - high performance pressure system 7

- - separation 6, 7
olive pulp 19–41
- phenolic compounds 19
Oryza sativa 46
ovoperoxidase 102

PAW (phenol-acetic acid-water) 52
peaches
- melting flesh 134–135
- non-melting flesh 134, 135
pectate
- hydrolysis 9
- lyase (PL) 169
pectic polysaccharides,
- NMR spectra 40
- - anion-exchange chromatography 32–33
- - in SDS extracts 41
pectin 2, 76
- after extraction with CDTA 27, 28
- HPLC separations 130–134
- methyl esterases (*see* PME) 165–177
- polymer protection, heat inactivation in precence of ethanol 51, 52
- structure 129–142, 168
PEG (polyethylene glycol) 64, 65, 69–70
- exclusion of 69–70
- long-term exclusion of 69, 70
peroxidase
- activity, staining 15
- isoenzymes 14–15
- - electrophoresis 14–15
- isoelectric focusing 14, 15
perturbation of adhesion 107–108
PG (polygalacturonase) 169
- endo-PG (endo-polygalacturonase; *see also there*) 134, 137–139, 141
pH fractionation, purification of xylan 34
Phaseolus 69, 70
phenol-acetic acid-water (PAW) 52
- isolation of cell walls 52
phenolic compounds in olive pulp 19
phenylpropenoid copolymers (lignins) 15–16
Phytophthora 171
Phytophthora cinnamomi 99, 106
Pinguicula 98
Pinus pinaster 46
Pisum sativum 46
PL (pectate lyase) 169
plant-substratum adhesives 95–110
plasma membrane
- isolation 185
- proteins, biochemical characterization 191–195
- vesicle fraction 185
plasmolysis 67, 68
PM (plasma membrane) 181
PME (pectin methyl esterases) 165–177
- detection of PME activities 165–167
- enzymatic properties 173
- fruit maturation, role of PME 176
- localization 167–168
- properties 169–175
- role as plant pathogens 175–176
- structures 174–175
PMIP (PM intransic protein) 182
polyethylene glycol (*see* PEG) 64, 65, 69, 70
polymers, hydrophobic 72
polyoxin inhibition, chitin synthases 82, 83
polypeptide depletion, callose synthase 187–188
polysaccharides
- anion-exchange chromatography, separation of polysaccharides 8
- cation exchange chromatography, fractionation of plant polysaccharides 9
- cell walls 1–12, 147–162
- - fractionation 29–34
- - graded precipitation with ethanol 30
- - ion exchange chromatography 7, 8
- - NMR studies 39
- *Ciucanu-Kerek* methylation 36, 37
- cross-linked pectic 29
- derivates, fractional precipitation 11–12
- fractionation by chromatography 5–10
- *Hakomori* methylation, polysaccharides 36
- hemicellulosic, anion-exchange chromatography 32–34
- insoluble affinity chromatography 10
- "loose" complex, interactions of carboxyl group of pectic polysaccharides and CDTA with calcium 27
- methylated polysaccharides, carboxyl-reduction of 37
- methylation analysis 36–39

- NCP (non-cellulosic polysaccharides) 2, 16, 74
- non-pectic 2
- pectic 32, 33, 40
- - anion-exchange chromatography 32, 33
- - NMR spectra 40
- - in SDS extracts 41
- precipitation 2–5
- - with inorganic salts 3–5
- - with ionic detergents 5
- - with organic solvents 2, 3
- size chromatography 5–7
- slime 77
pore
- size, definition 64
- structure, cell wall 76–77
porosity, altered, evidence 78
PR (pathogenesis-related proteins) 81
precipitation
- of glycoproteins 13
- polysaccharides 3–5
- - with inorganic salts 3–5
- - with ionic detergents 5
- - with organic solvents 2–3
- - reactions 2–5, 13
preparation of CWM, solobilization of material 23–24
primary wall 12
- porosity 66
product entrapment, callose synthase 191
pronase, effect on callose synthase 188
protease stimulation, chitin synthases 82, 84
proteins
- biotinylated 73
- in cell walls 12–15
- content AIR 21
- PR (pathogenesis-related proteins) 81
- proline-rich (PRP) 12
- uptake through pores 69
proteoliasin 102
PRP (proline-rich proteins) 12
Pseudomonas fluorescens 106, 171
purification of cell wall material 49–51

Raphanus 67, 69
rate of movement, F-dextrans 73
RG (rhamnogalakturonan) 129
rhamnogalakturonan (RG) 129
rhizoid tip, peroxidase secretion 101
rigidity of wall polymers 78
root cap cell walls 77

Saccharomyces cerevisiae 82, 83, 84, 90, 91, 104
Saeman hydrolysis conditions, glucose release from cellulose 34
Saprolegnia monoica 82, 84, 87, 88, 89
SDS extracts, pectic polysaccharides 41
SDS-PAGE (sodium dodecylsulphate-polyacrylamide gel electrophoresis) 14
secondary wall 12
sequential extraction, cell wall material 24–29
sidedness, membranes 185
size chromatography, polysaccharides 5–7
slime polysaccharides 77
sn-CR (supernatant of the dialysis of CR) 27
sodium dodecylsulphate-polyacrylamide gel electrophoresis (SDS-PAGE) 14
solute exclusion, pore size 67, 69
Sporobolomyces ruberrinus 46
sticky compounds 98
suberized walls 72
substrate adhesion, *Fucus* 100–103
subunit composition, callose synthetase 181–196
sugar composition
- alkali-soluble hemicellulose 8
- of cell wall
- - after anion-exchange chromatography 33
- - material after sequential extractions 28
- of fractions, after ethanol precipitation, from CDTA extract 31
- purified cell wall material 23
sugar oligomers, HPAEC 151

tagging adhesive compound, methods 107
TBS (tris-buffered saline) composition 194
Thalassiosira fluviatilis 88
thermospray interface, operating conditions 149
tissue
- firmness 135
- prints, extensin 121–122, 123

topology, callose synthase, probing 187–191
tracer
- molecules, functional assays' transport across cell wall 65, 66
- techniques, uptake of particles 69, 71–75
Tradescantia 69, 73
transmembrane bridge formation 102, 103

ultrastructural methods, wall porosity 64
ultrastyrogel columns 11
uptake molecules, tracer techniques 71–75
uronic acid, calorimetrical determination 35
Ustilago maydis 91

vanadate peroxidase 101, 103
vectorial proteolysis, callose synthase 188–189
Vigna
- *V. angularis* 46
- *V. radiata* 169, 170
visualization membranes by ECL 194–195
in vitro
- autolysis experiments 45
- cell wall
- - autolysis 45
- - loosening 45
vitronectin 102, 103
in vivo localization extensin 121

wall (cell wall)
- autolysis 45–57
- - autolysis in vitro 45
- binding enzymes 45
- bound enzymes 1
- components, fractionation 1–16
- composition during growth 78
- isolation
- - factors affecting 46–49
- - isolation by phenol-acetic acid-water 52
- - from olive pulp 19–24
- localized enzymes 76
- loosing, in vitro 45
- material (*see* CWM) 21 ff.
- matrix 70, 75
- movement
- - hindered movement 75
- - of macromolecules through 72–74
- polymers 19–41
- - major groups 37–39
- - sequential extraction 24–29
- polysaccharides 147–162
- - fractionation 29–34
- - NMR studies 39
- pore structure/porosity 63–78
- - alterated, evidence 78
- - microscopic visualization 66
- - variations of wall porosity 77–78
- preparations 117, 118
- proteins, purification 118–119
- rigidity of wall polymers 78
- root cap cell walls 77
- secondary wall 12
- sugars 28
western blots, extensin 121–122
wettability 105

X-ray diffraction analysis, chitin 86–87
xyloglucan fragments 10
xyloglucan-derived oligosaccharides, separation 6–7
xylose oligomers, HPAEC-MS 154

Zea mays 13, 14, 46, 74
zygote-bound adhesive material 101

MIX
Papier aus verantwortungsvollen Quellen
Paper from responsible sources
FSC® C105338

If you have any concerns about our products,
you can contact us on
ProductSafety@springernature.com

In case Publisher is established outside the EU,
the EU authorized representative is:
Springer Nature Customer Service Center GmbH
Europaplatz 3, 69115 Heidelberg, Germany

Printed by Libri Plureos GmbH
in Hamburg, Germany